RIDING FOR CAESAR

T0296699

RIDING FOR CAESAR

The Roman Emperors'
Horse Guards

MICHAEL P. SPEIDEL

HARVARD UNIVERSITY PRESS
Cambridge, Massachusetts · 1994

For Silvia and Markus
to remember their father.

Copyright © 1994 by Michael P. Speidel

Printed in the United States of America
Library of Congress Cataloging-in-Publication Data
Speidel, Michael
Riding for Caesar: The Roman emperors' horse guards, Michael Speidel.
        p.   cm.
Includes bibliographical references and index.
ISBN 0-674-76897-3
1. Guards troops—Rome.   2. Rome—Army—Cavalry.   I. Title.
U35.S648   1994
355.3'51'0937—dc20
93-23539
CIP

CONTENTS

PLATES

MAPS AND DRAWINGS

ACKNOWLEDGEMENTS

Archaeologists and Roman army scholars across Europe helped me in this undertaking, above all G. Alföldy, D. Baatz, H. Bellen, A. Birley, E. Birley, J. Bogaers, W. Boppert, P. Connolly, J.C.N. Coulston, H. Devijver, K. Fittschen, R. Friggeri, H. Gabelmann, L. Keppie, S. Panciera, M. Roxan, M. Sayar, B. Scardigli, M. Speidel the Younger, I. di Stefano Manzella, and H. Ubl. I thank them as well as J. Cooke and I. Newby who turned my phrases, B. Ikeda who drew the maps, and my wife Gisela who cherished my quest. Their place is with the happy ones (*Aeneid* 6, 663) who by their knowledge, skill, and deeds, made life a better thing.

PICTURE CREDITS

FOREWORD

Tall, fierce, and faithful, Caesar's German horsemen rode with him from battle to battle. When he became ruler, they became the imperial horse guard. The Republic allowed no guard in peacetime, yet one-man rule hinged on it – and Augustus, when he 'restored' the Republic, kept the horse guard. Praetorians and horsemen, the two branches of his guard, heralded the Empire.

The more the empire came to center around the ruler's person, the more the horse guard thrived. In 52 BC it numbered 400 troopers, under Trajan 1000, under Septimius Severus 2000, in the Late Empire 3500. Watching over the emperors' safety, the guardsmen had their hands on the reins of history. Courted by the rulers, they led the high life. As a training school and officers' academy they shaped the Roman army; as a shock force in the endless wars of the second and third century, they upheld the empire. And when in 312, in the battle at the Milvian Bridge, the guard went under, Rome ceased to be a capital of the Empire.

During these 360 years, the emperors' horse guard made its mark. Triumphal art on columns and arches highlights the horsemen so much that Dante, who loved Roman art, portrayed them in God's telling relief on Mount Purgatory:

> Here was shown the high glory
> of the prince of Rome
> – I speak of the emperor Trajan –
> horsemen thronged around him.

In our time an unmatched wealth of the horse guard's stone monuments has come to light, making the *equites singulares Augusti* the best-known Roman fighting force. In their forts in Rome, a magnificent series of altars was unearthed. Their graveyard at the catacomb of St Peter and Marcellinus has yielded hundreds of headstones, and from across the Tiber come the tall gravestones of the Julio-Claudian *Germani corporis custodes*.

Not long ago, further finds at Anazarbos in Cilicia gave proof that during the third century the horse guard still bore the name *Batavi* – as it had done in the first century. It follows that *Batavi, Germani corporis*

10

custodes, and *equites singulares Augusti* were but different names for the same unit. Now, for the first time, we can trace the history of the horse guard from Caesar to Constantine.

It is a striking story, for while the lure of antiquity lies above all in its majesty and sweeping panorama, the horse guard also beckons with rich detail. Its carved and inscribed stones tell much about the troopers themselves, about their homes on the northern frontiers, their forts in Rome, their weapons, training, and careers, their duties, families, and beliefs.

To ponder the looks, lives, and deeds of these men is not idle. Their build, their daring, skill, and faith awed even the great. Caesar praised them in his *Commentaries*, Trajan had them carved on his Column, Hadrian wrote poems about them. Two thousand years later, they strike us as men who matched the greatness of the empire and gave rise to the aristocracy of the Middle Ages.

M.P. Speidel
Mt. Tantalus, Honolulu
Fall 1993

1

FROM CAESAR TO NERO

Picked auxiliary horsemen, called Batavi *after the island in the Rhine, the finest horseback riders.*

Dio 55,24,7.

Caesar's *Germani* horsemen

By 52 BC, Caesar had brought the Gauls to their knees, yet in the winter of that year, under Vercingetorix, they rose in arms against him with such fury that all seemed lost. No sooner had Caesar brought his legions together than the vanguard of the Gaulish horse bore down on him at the town of Noviodunum. Drawing up his regular cavalry in battle line, Caesar sent it to meet the Gauls. Before long his men began to waiver. At that critical moment, when all was at stake, Caesar threw his *Germani* into the fray – 'some four hundred horsemen he had with him from the beginning'. The Gauls, unable to withstand their onslaught, broke and fled. Caesar's horse guard thus saved him from being trapped in certain defeat.

Holding back reserves until the decisive moment, Caesar had won by tactical skill. It is nevertheless astonishing that only four hundred men made such a difference. They must have been the kind of men Caesar's own army feared, 'huge, unbelievably bold and expert fighters'. Perhaps they attacked at a risky full gallop. Were they regular troops of the line or a guard? Caesar, saying that he had them 'with him', marks them as his escort, and his keeping them behind the battle line shows they were a reserve. As an escort and battlefield reserve, Caesar's German horsemen clearly were his guard. The history of the Roman emperors' horse guard thus begins at Noviodunum in Gaul in 52 BC.[1]

12

Where did Caesar hire these men? In 58, in the first year of the Gallic War, when he needed a trustworthy force to face the Suebian king Ariovistus and his horsemen at a parley, Caesar mounted the tenth legion on borrowed horses. Later that year he routed the Suebi and acquired German allies, the Ubians, who finished off the beaten Suebi. In 57, embassies from tribes east of the Rhine came with offers to join Caesar. Among them, no doubt, were Ubians whom he may have asked for horsemen since Roman field marshals, like Hellenistic dynasts, were wont to rely on foreign bodyguards. The Ubians ever after kept in touch with Caesar and, under the Julio-Claudian emperors, supplied more men to the emperor's horse guard than any other tribe save the Batavians. Very likely, therefore, Ubians served already in Caesar's guard.[2]

In 49 BC, when Caesar hastened to war in Spain, 900 horsemen rode with him, surely his German horse guard as well as Gaulish horsemen. After the battle of Pharsalus in 48 he took 800 horsemen of the guard along to Egypt, among them also Gauls. Caesar used the Gauls in his assault on Pharos island. Later, when he broke the siege at Alexandria – sweetened by Cleopatra – and faced the Egyptian army across the Nile, his German horsemen showed their mettle. They scattered and rode into the river to find fords for the legions. Where the banks were less steep, they swam across, established bridgeheads and routed the enemy, thereby allowing the legions to cross. Since Batavians later made up the bulk of the horse guard, and since they excelled in swimming rivers in full battle gear, Caesar's *equites Germani* on the Nile, too, may have been Batavians.

From Egypt, Caesar went to Asia, and on to Rome, traveling at great speed 'with horsemen in light order' – clearly his guard. Escort duty with Caesar was not for laggards. Keen horseman that he was, he dashed from one theater of war to another and demanded of his soldiers instant readiness, in any weather, at any time, day or night. He himself swam across rivers that would otherwise have hindered his lightning travels of up to a hundred miles a day. His guard had to ride as fast and swim as well.[3]

In the African war of 46, Caesar again had his German horsemen with him. When he faced Labienus, he found that his former second-in-command, now his opponent, also had a guard of Gaulish and German horsemen. Some of these were men of Caesar's army who had come to Africa with Curio in 48 BC. Spared after Curio's defeat, they had joined the Pompeians; now they made a truce with Caesar's guardsmen to see how they could avoid fighting each other. Labienus' horsemen, however, like Caesar's, balked at betraying their

13

trust, and the talks broke off. Afterwards, in battle, Labienus' other cavalry fled, but his Gaulish and German horsemen stood their ground. Attacked from above, and set upon in the rear, they were killed to a man. When the dead had been stripped, Caesar, going over the body-strewn field, was awed by their huge build and by the fact that, having been spared before, they wanted to return that favor by keeping faith with the Pompeians to the last. The German ideal of being 'true' – to repay favours with faith – was the same as the Roman ideal of *fides*.

The 'wonderful bodies' (*mirifica corpora*) that astounded Caesar are highlighted by a relief on the victory monument at Adamklissi. There, two warriors of staggering size are guarding Trajan who leans against a tree, the staff of command under his arm, as he watches a battle in the wooded hills of Dacia. No horses are shown but they must be nearby, for in the field the emperor always rode on horseback. The two warriors, therefore, are likely to be *Batavi* horsemen of Trajan's bodyguard. The geographer Strabo likewise saw Germans as larger and fiercer than Gauls, with redder hair.

From the anonymous author of the *African War* we learn that Caesar had raised his horsemen 'by influence, money, and promises' (*auctoritate, pretio, et pollicitationibus*). Caesar's influence may have brought him Gauls; money and promises brought him Germans. Promises must have meant cash awards once the wars were over. The wars never ended, but from here, surely, dates the custom that the emperor's horsemen, unlike other auxiliaries, could look forward to cash awards (*commoda*) upon discharge – which made them partners in the emperor's undertakings.

Labienus swelled the number of his Gaulish and German horsemen not only with prisoners of war, but also with half-breeds, freedmen, and slaves – all sorts of beholden men. This source of recruits for the guard was never wholly given up, and indeed some of Tiberius' guardsmen were former slaves. Slander made this a stain on the guard: Caracalla was rumored to have bought slaves in Germany for his *Lions*, and Galerius was said to have raised his horse guard from prisoners of war.

We are not told whether Caesar took his German horsemen on the last of his campaigns, the Spanish war of 45. Yet since he had taken them to all the other battlefields of the Civil War, and since he always needed a guard of fighting men, they are bound to have gone along, racing as best they could over the 2400 km (1500 miles) from Rome to Cordova in just 25 days, an average speed of 90 km (55 miles) a day. As for Caesar, he must have traveled in a carriage, for he found time to describe the journey in verses. No poems are likely to have been written by his horsemen during those 25 days, and most of them lagged behind: upon arrival Caesar found himself in want of a horse guard.[4]

Back in Rome, three months before he was murdered, Caesar took a pleasure trip to Campania, visiting Cicero in his villa at Puteoli. In a letter to a friend, Cicero frets that his villa was overrun by soldiers, nearly 2000 of them! The units to which these soldiers belonged are not known, but since Caesar never mentions a praetorian guard and probably had none, and since he traveled on horseback, some of the soldiers, surely, belonged to his horse guard. Later, passing the house of Dolabella, Caesar honored the Consul-elect by having his men parade to the left and the right of his horse – the first time we see the horse guard in its role of lending luster to the sight of a Roman ruler.

Caesar's German horsemen had served well as a crack battlefield unit and an escort. Shortly before the Ides of March in 44 BC he made them his sole bodyguard, dismissing his former Spanish guard. The Spaniards were armed mainly with swords and thus were not primarily horsemen. Although they had been utterly faithful, in the planned Parthian War Caesar needed first-class horsemen around him wherever he went. Augustus later likewise used his German horsemen as bodyguards, calling them *Germani corporis custodes*.

Caesar thus was the founder of the emperors' horse guard. Having carried the Roman frontier to the Rhine, he recruited his guard from the tribes there, setting the pattern for two hundred years to come. By forging the horse guard into an efficient fighting force, reading to ride with the ruler across the length and breadth of the empire, he laid one of the foundations of the new monarchy. The guard's tireless speed, its willingness to fight, and its trustworthiness never failed him. Its far-flung journeys in Caesar's service (Fig. 1) breathe a dogged steadfastness that matched the pace of the Genius and the vastness of the empire.[5]

Augustus' horse guard

After Caesar's death, his German horsemen sided with Octavian. In 43 BC they went over to Mark Antony, but so did Octavian, and both leaders of the Caesarian forces had German horsemen with them at the battle of Philippi in 42. Afterwards Mark Antony took some of them with him to the Orient, but Octavian, too, during his campaign in Sicily in 36, had a *Germani* guard, very likely the same men as Caesar's, for he would not have sent home battle-tried guardsmen

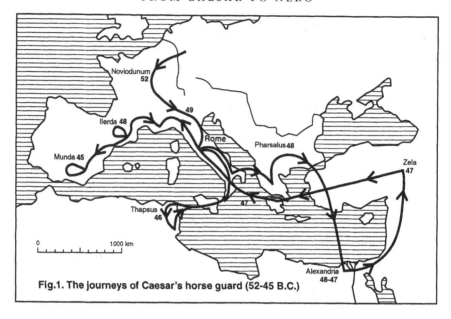

Fig.1. The journeys of Caesar's horse guard (52-45 B.C.)

only to ask, not much later, for new and untried troopers. To be sure, losses had to be made up, and veterans had to be replaced by recruits, but there is little doubt that the same unit guarded Caesar and Augustus. As under the later Julio-Claudian emperors, the troopers seem to have been Batavians and Ubians. Their service in the emperor's guard would explain why both tribes were admitted to the Roman side of the Rhine, the Ubians as early as 38 BC, 'for having proven their trustworthiness' – surely in the horse guard.[6]

Whether Augustus' bodyguards were all horsemen is not known, though the fact that they are called *Batavi*, like the third-century horse guard, suggests that they were. The emperor chose Batavians not for being foreigners, but for being the finest horsemen anywhere. His legate in Lower Germany no doubt picked them from tribal warriors who as allies had proven their horsemanship and fighting skill.

Our sources seldom mention Augustus' horse guard, since he, unlike his successors, wanted to keep so monarchical an institution in the background. In a sense, it was because of the guard that he had to forego the title *Romulus* he craved. Livy, whom Augustus called a

1 *Gravestone of Flavius Proclus, found at Mainz, Germany. Proclus, Arab archer of the horse guard, draws a composite bow, pulling the string with his middle and index fingers in the 'western release' towards his chest.*

16

FLAVIVS PROCVS
Q S ING AVG DOMO
O DELPIA AN XX

D · M ·
T AVREL SVMMVS EQ
SING · AVG · CLAVDIO
VIRVNO · NAT NORIC
VIX ANN XXVII · MIL
ANN · VIIII · P AELIVS
SEVERVS · PHERES
AMIC · OPTIM · F · C

Pompeian because of his political outspokenness, pointedly tells that Romulus was beloved 'more by the soldiers than by the senate', for 'he kept 300 armed *Celeres* ['Swift']-horsemen about him as a guard, not only in war but also in peace'. Indeed, Livy was not above broadcasting the wicked whisper that the senators had torn Romulus to pieces for being a tyrant.

Since archaic Greek times, so history told, tyrants had used their bodyguards to oppress the free, witness Peisistratos in Athens and Pausanias in Sparta. Caesar had claimed to be a new Romulus, and now when Octavian, too, wanted to be Romulus, Caesar's bloodied ghost stirred again. To avoid his adopted father's fate, Octavian had to settle for the title *Augustus* – all because of Romulus' horse guard. The bodyguard must have made Caesar and Augustus look like tyrants. Cicero's grumbling that Caesar came on a visit with 2000 armed men shows that by its numbers alone the guard overwhelmed even the most high-flying citizen.[7]

History had changed. Rome had become an empire, and the days of the citizen foot soldier drew to an end. Horsemen first held sway on battlefields in about 1000 BC in Assyria. From that time on, horse guards, elite household troops of the rulers, often 1000 strong, played a key role in the empires that followed one another. Sargon of Assyria in 714 BC boasted:

> With one thousand fierce horsemen, bearers of bow, shield, and lance, my brave warriors, trained for battle, who never leave me, either in a hostile or in a friendly land, I set out and took the field.

The Persian emperors likewise took to the field with 1000 picked horsemen, and so did the Seleucid kings in Syria as well as the Ptolemies in Egypt. Alexander, like Septimius Severus, had 2000 horse guardsmen, for to him they were a first-strike force. With the rise of the great marshals of the Late Republic, Rome fell into this pattern.

Unable to hide the empire's need for a ruler and the ruler's need for a guard, Augustus could claim, at least, that the fine state of the army, the long peace, and his own high standing freed him from the need to fawn upon the soldiers. He showed this by placing the fort of the horse guard north of the Tiber, far from the Forum and the Senate

2 Gravestone of a horse guardsman, Rome. The funeral banquet and the groom, long-reining the horse, are motifs brought from Lower Germany.

Hall (see Fig. 8), and by balancing the *Germani* bodyguards with a corps of truly Roman praetorian horsemen.[8]

Late in Augustus' reign, a certain Pusio heightened the horse guard's fame with an astonishing feat. As a 'German horseman', Pusio served the imperial princes in Dalmatia in AD 9, when Germanicus was thwarted by the walls of Splonum. Pusio hurled a rock of such size against the wall that it broke the breastwork and brought down a defender who was leaning against it. This so frightened the others that they fled and the wall was taken. Throwing rocks was not at all beneath elite warriors. In the Persian war, at Plataiai, Sparta's Aeimnestos knocked the enemy commander Mardonios off his horse with a well-flung rock. Roman and German horse guards alike hurled stones up to a pound in weight.

Pusio's story proves that horsemen of the guard, here as elsewhere, took part in sieges. It also bespeaks the troopers' huge size. The tallness of the hero of Splonum is borne out by his very name, handed down in two variations, Pusio and Pulio. *Pusio* is more likely, for during Augustus' reign, another giant, over ten feet tall, likewise bore the name of Pusio which means *Little One*. Both men were butts of the same joke that pitted their strapping size against their belittling name. The emperors needed stout guardsmen standing by, strong enough to overpower anyone.[9]

When in AD 9 news came of the crushing defeat of the Roman army in Germany's Teutoburg Forest, Augustus cashiered his bodyguard for fear that they might turn against him. While the guardsmen may not have been given to German nationalism, they nevertheless felt bound to their home tribes. Faithful as they were, the bond with their tribes was a weak point in their loyalty. Even the tribes west of the Rhine had wavered in their allegiance to Rome, and the guards may have listened to secret envoys from home. Indeed, they must have taken some steps that smacked of treason, for otherwise they would not have been banished. The danger they posed lay less in their number than in their being near the emperor's person. Augustus therefore sent his guardsmen to islands of exile.[10]

Tiberius

Five years later, at the beginning of Tiberius' reign, *Germani* body-guards are found again in Rome. Augustus had banished his guards when he was 72 years old. Perhaps he had no great need for them anymore, and we do not know whether he ever recalled them.

Tiberius, however, waging war in the Balkans and on the Rhine, needed not only a horse guard but every soldier the garrison of Rome could spare. The guardsmen Augustus banished thus were merely the few left in Rome after much of the unit had gone to the Illyrian war of AD 6–9. When in AD 10–12 Tiberius campaigned in Germany, he could fill any gaps in the ranks of his horse guard (whether imperial or princely) with Batavians and Ubians since those tribes had stayed loyal. Returning to Rome in AD 12, Tiberius must have brought his guard along, for it was there in AD 14. Emperors who took to the field, or who had to fear murder and mutiny, could not do without a horse guard.

The overriding concern of a Roman emperor was personal safety, *salus imperatoris*, besought on altars all over the empire. Next to safety came the loyalty of the armies, *fides exercituum*, untiringly proclaimed on coins. The horse guard, like the praetorians, answered both these needs: as guards they saw to the emperor's safety, while as under-officers and officers sent to the frontier armies, they underpinned the troops' loyalty to the emperor to whom they owed their promotion.

There was a political price to be paid, though, for bodyguards made an emperor look like a grim tyrant. Wags knew a tale of what befell a high-born senator under Tiberius: in asking the emperor's forgiveness for an untoward remark, he clasped Tiberius' knees and unwittingly caused him to fall. In a flash the dreaded bodyguards pounced upon the hapless senator and nearly killed him.[11]

Early in his career, in 9 BC, Tiberius learned what skill and steadfastness a horseman could offer. When his brother Drusus, had broken his leg and lay dying in the woods of Germany, Tiberius hurried in carriages day and night at breakneck speed from Italy across the Alps to Mainz. To see his brother before he would breathe his last, he hastened from Mainz on horseback in 24 hours over the last 320 km (200 miles) through newly conquered land. Changing horses as he went, the imperial prince accomplished the journey with only one companion, his faithful guide Namantabagius. Prince and horseman set a new speed record, a feat that went down in literature. Namanta-bagius' Celtic or German name, his skill in riding, and his role of escorting the prince show that he was a horseman of the guard. Certainly, he embodied the qualities the emperors looked for in their guardsmen. Small wonder that Tiberius kept such a guard even after Augustus dismissed his.

The horse guard's most dramatic mission came when the Pannonian legions rioted after the death of Augustus. To quell the uprising, Tiberius sent his son Drusus to Pannonia with two praetorian cohorts, strengthened with picked soldiers from other cohorts. Since horse throughout history served to discipline foot, Tiberius gave Drusus also the bulk of the *Germani* horse guard and most of the praetorian horse. The horsemen proved their worth as the emperor's trusty henchmen, guarding the prince amidst the mutinous legions and cutting down the boldest of the brawlers.[12]

In AD 16 Germanicus, Tiberius' adopted son, fought against the Cheruscans in Germany. In his battle-ready marching order four legions went before him and as many behind, while he made his way with two praetorian cohorts and 'picked horsemen'. The picked horsemen, no doubt, were *Germani corporis custodes*. Some of them may have been Germanicus' own private guards, but most, like the praetorian cohorts, no doubt belonged to the imperial guard. The guard troops on this campaign, then, were nearly the same as those who under Drusus had quelled the mutiny of the legions two years earlier. Here, as in the Dalmatian war of AD 9, the horse guard took to the field even though the emperor stayed in Rome.

Whether they went to war in person or not, the Julio-Claudian emperors needed first-rate horsemen as a key corps in their household forces. Their *corporis custodes* differed from the praetorian horse mainly in that they were horsemen of greater strength and skill. Juvenal, therefore, may refer to them rather than to praetorian horsemen, when he quips that, like Sejanus, anyone might wish for himself

> spears, and cohorts,
> outstanding horsemen, and a fort of his own.

Juvenal's *egregii equites* echo Dio's description of Augustus' bodyguard as the finest of horsemen.[13]

The scope of the guard's training may be gauged from what is known of the guard of King Maroboduus, ruler of the Marcomanni in Bohemia and one of Rome's sturdiest enemies. 'By ceaseless training', the Roman guard officer Velleius Paterculus asserts, Maroboduus had, 'in a short time, brought the discipline of his horse guard to such a pitch that it came close to Roman standards, threatening even the Roman empire.' In Maroboduus' army the horse guard clearly was the strategic core and it differed from the rest above all by the quality of its training. The same was true for the Roman army, in that the praetorians and the horse guard together constituted its strategic core and its most highly trained elite corps, under Augustus and Tiberius as well as under Hadrian and in later centuries.[14]

Caligula

Under Caligula the horse guard took center stage. The great traumatic experience of Caligula's youth was the revolt of the legions upon Augustus' death. He was two years old in AD 14 when rebellious legionaries held him and his mother hostage and nearly drove his father Germanicus to kill himself. In those dark hours the horse guard kept faith and sided with Germanicus. Caligula remembered this – or rather tales of it – all his life, and felt safe only in the midst of a strong horse guard. As a prince he must have had 'private' *Germani* guards like his brothers. As emperor he indulged his horse guard, beguiling the men with rich gifts. This was the way German kings treated their guards, and it was the pattern Caesar had set when he raised his horsemen 'with pay and promises'.

With his horsemen Caligula charged across the fantastic pontoon bridge he built at Puteoli in 39. The bridge, some 5 km (3 miles) long, spanned the bay to the west of Naples from Baiae to Puteoli. It was a major engineering feat – its construction took so many ships that the grain supply failed and Italy went hungry. When it was ready, Caligula crossed the bridge on horseback, splendidly decked out in a cuirass, a golden coat, and a crown of oak leaves. At the head of a large cavalry battalion – the *Germani*, no doubt, as well as praetorian horsemen – he broke into a gallop near the end of the bridge and, with shield and sword, wildly charged into Puteoli as if in hot pursuit of a foe. After resting there 'from battle' for a day, he rode back in triumph the next evening, standing on a chariot, followed by the praetorians, loaded with 'booty' and 'prisoners of war'. Before dinner, he harangued the soldiers and rewarded them with gifts of money.

All this was clearly a military maneuver by the guard, the empire's strike force. Suetonius says it was meant to overawe the tribes of Germany and Britain, against whom war preparations were then in full swing. The dramatic high point, the emperor's charge at the head of his horse guard, should be discounted, perhaps, as merely symbolic, like Trajan's charge in the great relief now on Constantine's arch, or the stock image on coins showing emperors riding down an enemy warrior. The hoped-for strategic effect of the guard's maneuver, however, is borne out by a striking parallel: Hadrian's horse guard, swimming across the Danube, frightened the enemy tribes into submission. Caligula showed the guard's readiness for a sudden strike by the engineering feat of the bridge, built, no doubt by the praetorians but crossed by the whole guard, both foot and horse. It proved the mettle of Caligula's household forces, for the *disciplina* shown in building a bridge bespoke the true quality of an army.[15]

That same year Caligula had to deal with a conspiracy. High-placed schemers tried to lure the legions of Upper Germany into a rebellion. Since he needed a pretext to go there without alerting the traitors, Caligula prompted the Clitumnus oracle to tell him he should go to Germany and bring his *Batavi* horse guard up to strength. That an emperor traveled in person to the land from which he sought recruits was unheard of. Yet people must have been willing to believe such a story, or there would have been no point in giving it out. Everyone, it seems, understood how much the horse guard mattered to the emperor. Rumor even had it that the huge war preparations then underway against Germany were undertaken just to replenish the guard. And indeed, once he had dealt with the conspiracy, Caligula went to the lower Rhine, perhaps to hire horsemen for his guard as if this had been his goal all along.

The second-century horse guard was 1000 men strong with some seventy new recruits taken in every year. For the strength of the *Germani corporis custodes* and their recruitment in the first century, however, we have no reliable information. Of Caesar's 800 horsemen in 47 BC some were Gauls, hence his *Germani corporis custodes* numbered perhaps as few as 400, as in the battle at Noviodunum. Yet the fact that Josephus calls their commanding officer in 41 a *chiliarchos*, commander of 1000 men, suggests a much higher strength for the time of Caligula. In the last years of Tiberius' reign their number may have dwindled so that Caligula had to bring it back up. In so doing he very likely swelled their ranks far beyond their former strength – that, at least, would explain the great lengths to which he went in enlisting new recruits and it would give meaning to Aurelius Victor's charge that Caligula brought foreigners and barbarians into the army as men who abided his tyranny. The more an emperor had to fear uprisings, the larger a foreign horse guard he needed.[16]

Whimsically staged field maneuvers with the horse guard made Caligula the butt of unfortunate jokes. Suetonius tells a story of how the emperor once ordered a few of the *Germani* guards to cross the Rhine and hide there. Then, when messengers came – after lunch – breathlessly reporting that the 'enemy' drew near, Caligula rushed with his friends and some of the praetorian horse into a nearby wood where they chopped branches from trees and dressed the trunks as trophies. Returning by torchlight, Caligula upbraided as feeble cowards those who had failed to follow him. To those who had shared his 'victory', he awarded newfangled crowns emblazoned with the sun for reconnoitering by day, with the moon and the stars for scouting by night.

The event makes sense only if one takes into account the swimming skills of the *Germani*, who were trained to cross the Rhine in full battle gear. It seems that – after lunch – those on the far side posed as the enemy and threateningly swam over to the Roman bank. When word came of their crossing, the emperor and his friends took to their horses, headed the 'enemy' off, and put them to flight. No doubt the 'enemy' *Germani* got away by scrambling into the river and swimming off in a hurry. There is no need to assume, as scholars have done, that the emperor crossed the river. However, there must have been some mock-fighting at least, otherwise Caligula could not justify putting up trophies bedecked with the weapons of fallen enemies to mark the spot where they turned and fled.

It was a maneuver well within the bounds of Roman military tradition. Caesar, as well as Trajan, more than once had his scouts give false alarm in order to keep the soldiers on their toes. And while guarding the emperor in the field should have been the task of the *Germani* bodyguard, Caligula here had the praetorian horse with him, for the *Germani* would have recognized the 'enemy' as colleagues.

The decorations Caligula handed out show that he wanted the action to be seen as a reconnaissance-in-strength. This was not a grossly overambitious interpretation to put on such an outing – although he may have kept up the shamming overly long. The ridicule that greeted him was a long-range political blow, for it made light of his maneuvers at Puteoli and on the Channel coast as well.

Once he had strengthened his horse guard, Caligula wanted to decimate the legions – kill every tenth man – to get even with them for beleaguering his father and himself 26 years earlier. He called the legionaries to an assembly without their weapons, even without their swords, and then brought up the armed horsemen. However, when he saw some wary legionaries fetching their weapons anyway, he gave up the idea and hurried back to Rome.[17]

Caligula appointed slaves and gladiators as commanders of his horse guard. Not only did he like such men around him – all 'bad' emperors did – but with their fighting skills they could also train the troopers. He entrusted several Thracian-type gladiators with the command and made Helicon, a slave, *archisomatophylax*, head of the body guard 'on the palace staff'.[18]

The day Caligula was murdered, the gladiator Sabinus had the command. Hearing that the emperor had been killed, the detachment on duty at the palace went berserk. Josephus says:

News of the death of Gaius first reached the *Germani*. They were the emperor's guard and were named the *Germani* unit after the nation from which they were recruited. It is a national trait of theirs to be carried off by emotions more than other barbarians, because they reason less about what they do. Their bodies are strong and they achieve much at the first onrush against those they see as enemies. They were shocked when they learned of Caligula's murder, for they judge not by what is good for all but by what is good for them. They had liked Caligula particularly well, for he had bought their goodwill with gifts of money. With drawn swords they burst out from the palace to find the emperor's murderers, led by their *chiliarchos* Sabinus who owed his command over these men not to the manliness and nobility of his ancestors, for he was a gladiator, but to his bodily strength.

The guardsmen went on, in Josephus' version, to kill various senators and to hold hostage, at sword point, the whole theater on the Palatine, teeming as it was with nobility. Thus they fulfilled their sworn duty: to prevent, or else to avenge, the murder of the emperor. Neither their oath nor their own sense of duty left them a choice. So dreadful was their rioting that some called it an uprising – clearly there were enough of Caligula's *Germani* around to make the city shiver.[19]

The 'sudden and quick in quarrel' looks of the *Batavi* led Martial to write a poem mocking people's fear of them. Inscribed on a clay mask, it read:

Sum figuli lusus, russi persona Batavi
Quae tu derides, haec timet ora puer.

I am a potter's jest, the mask of a red-haired Batavian.
Though you make fun of it, a boy fears this face.

Just such a mask, with blond hair tied in a knot on the right forehead, and with red war-paint, has been found in Italy. With its wide eyes, massive hair, and flappy ears, its grimace, mustache, and beard, the mask looks fearsome enough. Like the poem, mixing dread with derision, the mask may portray a tribesman of the terrible Batavian uprising in 69 or else a soldier in the emperor's horse guard – or both. It betrays the feelings with which city-Romans looked at the emperor's 'foreign' horsemen. It would be interesting to know whether the *Batavi* horsemen purposely kept up their outlandish looks or softened them for the city; the smart well-groomed looks of the second-century guardsmen (Plate 11) point to the latter.[20]

24

Claudius

Popular outrage insisted that Sabinus, the gladiator in command of the horse guard at the palace, pay for the killings after Caligula's death. Claudius agreed and wanted him done to death in a befitting way – in gladiatorial combat in the arena. Sabinus, however, of winsome 'bodily strength' was a lover of the new empress, Messalina. To everyone's chagrin, she prevailed over henpecked Claudius and Sabinus was spared.

Caligula had entrusted the command of the horse guard to slaves and gladiators like Sabinus. Claudius, on the other hand, favored imperial freedmen for high office and hence gave the command to a freedman. Nero, who wanted the horsemen to be well-trained for mock battles in the circus, again appointed gladiators.[21]

Caligula had blazed a trail for Claudius and Nero. Like Caligula, both gave the *Germani* rich gifts to buy their loyalty. Gifts, however, merely bought time, for they undermined the troops' discipline. If afterwards they were ever withheld, the mood would turn sour – a rule of social behaviour Roman field marshals knew very well.

The guard's new-found wealth during the reign of Claudius and Nero leaps to the eye from the tall headstones for horsemen who died during their service in Rome. Their huge Travertine limestone slabs – marble was not yet in fashion – have been found in the two graveyards of the *Germani* across the Tiber. They rival in size and fine lettering the gravestones of the praetorians, proclaiming the horse guard's equal status. At first the *Germani corporis custodes* had used small, poor man's plaques for gravemarkers and hence had no ancestral model to turn to once they became rich. Unlike Trajan's *equites singulares Augusti* who brought their own pattern from Germany, the *Germani corporis custodes* borrowed their gravestone style from the praetorians. The tallest of these headstones, belonging to one Gamus, towers at 2.2 m (7 ft), overtopping most of the praetorian stones. And though a wreath is carved in the pediment, alluding to immortality, and a sphere, hinting at the soul's journey through the heavens, the size of the slab rather than its decoration was designed to awe the passer-by.[22]

Gamus, like others, bears a Greek name, his 'brother' Hospes a Latin one. Greek names, often given to slaves or freedmen, suggest that in the beginning the horse guard comprised slaves as well as free men. Once established, the way the unit entered names in its rolls stayed the same, even if no slaves were enrolled any more. Gamus lacks Nero's name Ti. Claudius, hence he died before the Pisonian conspiracy in 65 when Nero bestowed his name and perhaps citizenship on the horse guard for having done his bidding unflinchingly.

From the inscriptions on the gravestones much can be learned about the guard itself. The full name of the Julio-Claudian horse guard was *Germani corporis custodes*, and so Suetonius twice refers to them. On the stones thus far recovered, however, the guardsmen call themselves *corporis custos* or *Caesaris Augusti corporis custos* (no less than 18 times), flaunting the title of the emperor in the name of their unit – just like the *speculatores Augusti* (guards), the *frumentarii Augusti* (spies), and later the *equites singulares Augusti*. Four times horsemen call themselves *Germanus*, which came to mean 'bodyguard'. Surprisingly, however, they never use the unit's less formal title *Batavi* (and neither did later guardsmen on their gravestones in Rome).

The commander's title was *curator Germanorum*. *Curator* in this sense was not a rank but the designation of an *ad hoc* commander. Troop leaders, often non-Germans, perhaps gladiators, had the title decurion, and the chief decurion seems to have served as the *curator*. It is astonishing to see that the guardsmen had a brotherhood, a *collegium*, from whose fees they paid for their gravestones, for regular soldiers were not allowed to have such fellowships lest they become hotbeds of crime and treason. The horse guard, even its common soldiers, no doubt were thought to be above such folly.[23]

The most dramatic life of any guardsman surely was that of Arminius' nephew Italicus. In the battle of the Teutoburg Forest in AD 9, Arminius, chief of the Cherusci, freed Germany east of the Rhine, while his brother Flavus kept on serving in the Roman army as a highly-decorated scout. Seven years later the two brothers waged a great shouting match across the river Weser, with Arminius sneering at the vile trinkets Flavus had to show for a lost eye and for betraying his own people. Flavus and a Chattian princess later lived as citizens in Rome where she bore him a handsome son with the telling name of Italicus. In 47 the Cherusci asked Claudius for the young man to become their king.

Italicus is said to have been skilled in Roman as well as in German fighting techniques and horsemanship. German skills entailed swimming rivers in formation with horses and full battle gear, while Roman tactics stressed intricate wheeling, even to the left, during attack. In Rome all this could be learned in only one place, in the horse guard, in which Italicus therefore must have held some rank. The emperors' horse guard needed to know Roman tactics to fight alongside Roman cavalry, but it also needed to know German tactics to swim across rivers – a skill of great value that had made the guard stand out with Caesar on the Nile, Caligula on the Rhine, Hadrian on the Danube, and Maxentius on the Tiber.

High-born foreign princes in the horse guard were welcome ornaments at court, and in the second century a Parthian nobleman is known to have been in its ranks. Nevertheless, it was astounding that Italicus as a horseman of the guard should become king of the Cheruscans, the very nation that had defeated the Romans some thirty years earlier under Arminius. Claudius not only granted the Cheruscans their wish but also gave Italicus money and bodyguards, perhaps some fellow *Germani corporis custodes*, to steady him in his pro-Roman stance. Italicus at first was much beloved by the Cheruscans, for he drank and caroused as they, but when success made him overbearing, they called him a Roman stooge, the son of a Roman spy, and drove him out. He was restored with the help of the neighboring Lombards. For Rome's policy towards Germany, Italicus had served well, and the same may have been true of other foreign princes among the horse guard.[24]

Nero

In the field the emperor rode on horseback. In peacetime he therefore trained and paraded with his horse guard. A coin with the legend *decursio* portrays Nero wearing helmet and cuirass as he rides with his staff of command ahead of a guardsman who holds the imperial standard. A *decursio* could be a true military training maneuver or merely a commemorative pageant, but in either case it was performed in full view of the soldiers and the public. It called for keen horsemanship, deft handling of weapons, and thorough training by the emperor with his guard. The *decursio* thus not only raised the emperor's skill in the field, it also showed him to the people fulfilling his role: on the parade ground, as on coins, he was manifestly the army's true leader.[25]

The horse guard also needed a commander of its own. Its paramilitary status in the Julio-Claudian period (their squadrons were servile *decuriae* instead of military *turmae*) made it easy for the emperors to appoint a commander from outside the army. Free to choose whomever they pleased, oppressive emperors, hated by the upper classes, could pick freedmen or gladiators who owed them everything and hence would back them unflinchingly, knowing they would fall if the emperor fell. By contrast, a praetorian career officer as commander might hatch plots even against a good emperor – there was no relying on Seneca's claim that a ruler bestowing good things on everyone was safe, needed no guards, and bore weapons just for show. For the first century we know only freedmen-gladiators and none of the regular commanders who may well have been true army officers.

Nero's commander of the *corporis custodes* was a gladiator, in part because of the horse guard's need for intensive training both for actual fighting and for ceremonies or mock battles in the circus. Even some troop leaders (decurions) were freedmen-gladiators. Emperors like Caligula, Nero, and Domitian stressed training for shows. Their guardsmen learned skills that might be taught by 'a little Greek rather than by a highly decorated Roman veteran', frowned the upright Pliny. Yet to give shows at the circus was a major duty of the emperor, and here Nero used the guard to the full. In what to us seems an appalling slaughter, his horse guard, in one show alone, speared to death 400 bears and 300 lions.[26]

To be guarded by tall, handsome horsemen heightened one's splendor and status. Nero honored his mother by assigning her a horse guard detail. On the dark side, an oppressive ruler might rely on the fear his guardsmen spread and use them more as henchmen than horsemen. In the Pisonian conspiracy of 65, the horse guard was Nero's willing helper in rounding up an endless line of suspects. Where regular troops balked at dragging so many high-born citizens to their deaths, Nero sent raw recruits instead, or *Germani*, whom he trusted, Tacitus says, 'as foreigners' (*externi*). Living in Roman Germany, the Batavians had long ceased to be the foreigners they once were. Ruthlessness, however, in Tacitus' view, was barbaric and foreign, so much so that he even charged Vitellius' Roman troops with 'foreign behaviour'. Tacitus' view of 'foreign' guards betrays itself in his finding fault with the Parthian king's bodyguard: 'All drummed out from their homelands, they didn't understand what was right, or mind what was wrong, being paid henchmen for crimes.' Truly Roman bodyguards, one is to infer, would not put up with a harsh, bloodthirsty tyrant.

The praetorians, for their services in the Pisonian conspiracy, received each 500 denarii – two-thirds of a year's pay – and, as a unit, free grain forever. The horse guard, for its share in the same dirty work, surely, earned no less. Second-century *equites singulares Augusti*, like the praetorians, lacked the rank of *actarius* who in the frontier units kept the accounting books for food. It seems, therefore, that both branches of the guard were awarded the privilege of free grain at the same time. Nero also gave the horse guard Roman citizenship, as can be seen from a gravestone mentioning three guardsmen all bearing the name Tiberius Claudius. The fact that they now bore the emperor's family name whereas before they had only individual names, proves that they had risen in status. In this, too, they were forerunners of the second-century *equites singulares Augusti* who bore the emperor's name.[27]

The role of the horse guard during Nero's last years is exemplified by the story of its commander, the gladiator and freedman Ti. Claudius Spiculus. Nero gave Spiculus the fortunes and houses of men who had celebrated triumphs, thereby raising him to the highest social rank a man of his beginnings could reach. Spiculus requited him well. When Nero's reign had become unbearable and Galba rose up against him, the senate somehow brought the praetorians to abandon Nero and to withdraw their guard detail. Spiculus, along with the horse guard, however, would not budge. Only when the *Germani* at last handed over their unyielding commander to the senate party and withdrew their guard as well, was Nero's fate decided. In his last hour Nero still called for Spiculus to give him the deathblow.

Spiculus held out for Nero to the end and was lynched as one of 'Nero's men'. The mob threw him on the ground at the Forum and tore down Nero's statues so they would crash on top of him. Why the *Germani* in the end handed Spiculus over to the anti-Neronian party is not fully known; very likely their faith had been shattered by the news that the army in Lower Germany had revolted. Historians later hailed them as the most faithful of guards.[28]

Sent home by Galba

It had not been easy to talk the guard into turning against Nero, and the *Germani* proved even more steadfast than the praetorians. It is unlikely, therefore, that Galba cashiered the *Germani* as betrayers of Nero, all the more so since he, too, was a 'traitor' and thus could hardly blame others for shifting away from the oppressor and towards the sovereign senate.[29]

Galba, in blind pursuit of the harshness of former times, boasted that he was wont to choose, not buy, his soldiers. Soon the guard was chafing under his niggardliness, so unlike the largess of Caligula, Claudius, and Nero. Whether their ill will led to any action is unknown. It had always been to the guard's great advantage to keep faith with the emperors, and Suetonius later said the horse guard had been true in case after case. Galba, however, claiming they supported the designs of Cnaius Dolabella, whose gardens across the Tiber abutted their camp, cashiered the whole unit and sent them home.

Cashiering a unit whose members had joined a conspiracy, was prescribed by law, as was the banishment from Rome of soldiers discharged in shame. However, ousting the horse guard also saved Galba a lot of money, and this seems to have been his true reason for disbanding the unit. He sent the men back without any cash award although, ever since Caesar hired his horse guard by pay and 'promises', they had a right to a handsome cash award at the end of their service. Galba paid for this. When his own murderers drew near, the men around him, such as he had, ran away and left him in the lurch. Suetonius, perhaps ironically, found it 'rather strange' that no one lifted a finger for Galba. The empire, too, paid a price out of all keeping, for it underwent a new round of civil wars.

Caligula's rich rewards had set the trend that the horse guard would be faithful only so long as it was well paid. The emperors of the next two centuries understood this perfectly well, as the magnificent gravestones of their horse guard prove. To be sure, the rest of the army also wanted its discipline and love of country to be underpinned by cash, but failure to pay the guard properly was asking for murder and mayhem.[30]

From 69 to 98 nothing is heard of the *Germani corporis custodes*. No gravestones and no literary reports tell us whether the Flavian emperors had a horse guard in Rome. The argument from silence has some weight in this case, since the accounts of Tacitus, Suetonius, and Dio for the period are partly preserved and might have mentioned such a guard had it existed. Likewise the many gravestones of *Germani* before 69 and of *equites singulares* after 98 stand in stark contrast to these years, when horse guard gravestones are altogether lacking.

Vespasian, fully backed by the frontier legions, could do with a much smaller household force. Vitellius had inflated the number of praetorian cohorts which Vespasian then cut by almost half, from 16 to 9. Nevertheless, for personal protection and pomp he, like all emperors, needed horse guardsmen. He could not revive the *Batavi*, though, for at the outset of his reign he had to fight a full-scale war against their tribe. 'Batavi guards,' in a pun by the poet Juvenal, now meant the opposite, namely Roman legions guarding the conquered Batavians:

> . . . *domitique Batavi*
> *custodes aquilas* . . .

The legionary eagles guarding
the beaten Batavian.

Vespasian may, or may not, have brought to Rome some of his *singulares* guards whom he had as a field commander in the Orient. Perhaps he merged them with the praetorian horse. Since we know so little about the horse guard from 69 to 98, it seems worthwhile to see in what way the praetorian horse and the emperors' *speculatores* could serve as a horse guard during the later first century.[31]

The praetorian horse

At least 400, and perhaps 1000, of the 10,000 praetorians served as horsemen. Like legionary horsemen they did not belong to cavalry troops (*turmae*) but to infantry *centuriae*, and they lacked a regular commander as well as the basic cavalry under-officers *signifer, sesquiplicarius, duplicarius*, and decurion. Like the legionary horse they could nevertheless fight as true cavalry and were not just orderlies, messengers, or guards for praetorian officers. They paralleled the horse guard in surprisingly many ways.[32]

The duties of the praetorian horse in Rome were manifold. For urgent errands, horsemen were better than foot soldiers since they clattered more quickly through crowded streets and, mounted, saw farther. Praetorian horsemen, therefore, as well as *Germani corporis custodes*, rounded up suspects in the conspiracy of Piso against Nero in 65. Praetorian horsemen killed Galba at Otho's behest – when they saw the emperor from afar, they rode over to him, scattering the crowd, and struck him down. Praetorian horsemen likewise hunted down and killed Nero.

To the people, the bread and circuses furnished by an emperor marked his success or failure. The praetorian horsemen therefore had to shoulder their share in public shows. Claudius had them and their tribunes, and even the praetorian prefect, kill African beasts in the circus. They also served in the sea battle on the Fucine Lake, the show-to-dwarf-all-shows staged by Claudius in 52. In a move modern criminal justice can only dream of, Claudius had 19,000 inmates from death row fight each other from battleships. Hoping to win a reprieve, they fought with great skill and pluck before a huge and eager crowd on the shore, among which Claudius and Messalina stood out, gorgeously dressed. The security problem posed by 19,000 armed desperados was perhaps not as overwhelming as might be thought, for the rowers were chained to their benches so they would do the utmost to save their ship rather than go down with her. Still, measures had to be taken to keep vast numbers of thugs from trying to break out and flee. Hence rafts, manned by praetorians, drew a huge circle around the battle site, taking care to leave enough space for the ships to gather speed and maneuver. The praetorians, infantry and horsemen if we believe Tacitus, stood by on the rafts behind ramparts, ready to fire catapults and *ballistae*. If the horsemen rode along the circle of the rafts, it was an awesome show indeed.

Praetorian horsemen strengthened the guard, making it a more all-round fighting force. The task force sent to quell the revolt of the Pannonian legions in AD 14 thus included not only two praetorian cohorts and the horse guard but also the praetorian horse. And they, too, guarded the emperor in the field. As shown above, during his campaign on the Rhine in 39, after Caligula had sent some of his *Germani* guard across the river to play 'enemies', he went reconnoitering with the praetorian horse. Clearly he had a choice between the two corps for his guard. No doubt this was to goad them into competition with one another, just as provincial field commanders had a double guard of *equites singulares* and *equites legionis*.

In the civil war of 69, when the *Germani corporis custodes* had been cashiered, the praetorian horse, taking to the field for Otho, fought at Cremona brigaded with 'auxiliary cavalry'. Like a true horse guard they were held in reserve, but once launched fought fiercely and keenly.[33]

Under Vitellius, many horsemen from the frontier *alae* became praetorians. Those that stayed on to serve the Flavian emperors kept the marriage rights they had as auxiliaries: they were allowed to have a semi-legal wife and, upon discharge, to legalize both marriage and children, while regular praetorians, serving only 16 instead of 25 years, were not allowed to marry. Horsemen transferred from the *alae* gave the praetorian horse of the Flavian period something like the fighting power of the earlier *Germani corporis custodes* or the later *equites singulares Augusti*, and over the years Vespasian may have brought in new men from the frontier *alae* to uphold the quality of his horse guard.[34]

In 89, a field detachment of praetorian horsemen fought gallantly against the Marcomanni, in line with Domitian's greater use of the guard in frontier wars. The praetorian horse of the time, recruited in part from the frontier *alae*, fulfilled the role of the emperor's horse

3 Marcus Aurelius reviews the horse guard: scene from the Aurelian Column in Rome. The troopers, smartly aligned to the left of their stallions, are holding staffs without blades – perhaps they are the hastiliarii-*escort. Behind Marcus rides the tribune of the guard, wearing a bell-shaped officer's helmet.*

guard both in the city and in the field. Their campaign strength at the beginning of the second century, reckoned by Hyginus to nearly match that of the *equites singulares Augusti*, stands as the high-water mark of their battlefield use. In the third century, however, they still took to the field with the emperors: one of their training officers died with Septimius Severus in Britain, and as late as 293 the praetorian horse rivalled the *equites singulares* with Galerius in Egypt.

In Domitian's war against the Marcomanni the commander of the praetorian horse was M. Arruntius Claudianus, a former prefect of an *ala* and thus a cavalry officer of the line. He earned five battle awards, more than any other equestrian officer had ever received or would receive for a long time to come. Such awards befitted an urban or praetorian tribune of the guard rather than an officer of the line. Arruntius' awards foretokened things to come, and indeed, within ten years, commanders of the *equites singulares Augusti* ranked equal to urban tribunes. Arruntius' gallantry and Domitian's acknowledgement of it by treating Arruntius as a tribune of the guard thus paved the way for the high rank of the horse guard commanders in the future.[35]

Speculatores guards

During the first century, elite praetorian horsemen formed a bodyguard, the *speculatores Augusti*, whose special weapon was the *lancea*. They escorted the emperors through street crowds, stood behind them at banquets, and could also hold their own on the battlefield. The name of the *speculatores* betrays their origin: reconnaissance was so essential to Roman field marshals, and so risky, that their reconnoitering force became their bodyguard.

The *lancea* of the *speculatores* has become known from a gravestone relief: its long shaft ended in a knob, and its short, broad, heart-shaped blade had a cross bar. Both ends, therefore, were designed for crowd control. This throws light on a dramatic incident foreshadowing Galba's murder by his own praetorian horsemen: once when a crowd suddenly surged forward, the spear of a *speculator* nearly wounded the emperor as he alighted from his horse. What had happened? A soldier

4 *Marcus Aurelius (his head replaced by that of Constantine) meets the horse guard as they set out to war: a panel, set in the arch of Constantine. The tribune offers the emperor the imperial banner, while three helmeted and mailed horsemen stand behind. Below beckons the Flaminian Way.*

would never wittingly aim his lance at the emperor. But if he used the butt of his lance against an unruly crowd to clear the way, the blade of the weapon might indeed point toward the ruler, all the more since Roman horsemen rarely tucked lances under their arms but held them low with straight arms. Then, when the crowd pressed closer, perhaps to get a better look at the emperor, it may have pushed the *speculator* back, driving the sharp end of his lance towards Galba.

Galba's mishap shows that one of the main duties of the *speculatores* was clearing a path for the emperor through crowds. Using non-lethal spear butts for this purpose was nothing new, they had always served to enforce discipline. Only when things got out of hand – or under tyrant emperors – would guardsmen use the sharp end of their lances. Suetonius lists Galba's incident among the signs portending the emperor's fall, and well he might, for a few months later, praetorian horsemen, perhaps the very *speculatores* of this episode, cantered over to Galba and cut him down.[36]

Though their precise number is unknown, the *speculatores* can hardly have been fewer than the 300 they have long been thought to be. Had they been fewer, they would not have counted in the field. Of Otho's withdrawal before the battle of Bedriacum in 69 Tacitus says: 'With him left a strong force of praetorians, *speculatores*, and horsemen – and those who stayed lost heart.' To be singled out like this, the *speculatores* must have been a sizable force. Three hundred seems to have been a good number for such an escort, since the *Celeres* horse guard of Romulus as well as the praetorian horse of the Republic and the Late Roman *excubitores* were all 300 strong, as were Alexander's Royal Troop and Napoleon's Mamluks.

A high-ranking *centurio exercitator* was drillmaster of the *speculatores*, hence there must have been more to their job than merely keeping watch in the palace, carrying messages, and standing by while the emperor dined. An *eques speculator* even received battle decorations. Clearly, their skills in riding, fighting, escort duty, and parading, were as well honed as those of the *equites singulares Augusti* in the second and third centuries.[37]

Like the *equites singulares Augusti*, the *speculatores* of the early first century were also called *Augusti*, 'the emperor's own', to set them apart from the *speculatores* of provincial armies. Sometime after 23, when all praetorian cohorts had been brought to Rome, the *speculatores* were fully incorporated into the cohorts. Thereafter they are no longer called 'the emperor's own', for being praetorians they were as much the emperor's men as all the others.[38]

Why were the emperors, among the crowds of the city and at

banquets, guarded by *speculatores* rather than by *Germani*? The rulers, it seems, did not want to unleash 'foreigners' against citizens lest they be seen as tyrants relying on ruthless barbarians. Tacitus' remark that Nero trusted the *Germani* because they were foreigners shows that such charges were bandied about. In 69 when former fleet soldiers besought Galba to confirm their legionary status, the emperor called in horsemen to scatter and decimate them. The horsemen may have been *Germani corporis custodes* or praetorian *speculatores*, but more likely the latter, unless Galba wanted to add insult to injury. Horse had always been used to discipline foot, but *speculatores*, being older men, may have taken greater care in their duties towards the emperor, towards other soldiers, and towards the people.[39]

Otho's rise to power shows what a few horsemen of the guard could accomplish when they served as the emperor's security police. He began his grab for power with no more than two *speculatores* and was proclaimed emperor by only 23 of them – so few could capture the praetorian camp, the capital, and the empire! During the civil war the praetorians, and surely also their *speculatores*, joined the winning side early enough to be in the good graces of the Flavian emperors. Hence the *speculatores* remained a special guard of the emperors, so much so that diplomas of 73 and 76 mention them as a unit of their own. Very likely they still served Vespasian, Titus, and Domitian as escorts. In 97, because of their treason against Nerva, they lost their role as the emperor's bodyguard. In Hyginus' account of the emperor's field army, written, perhaps under Trajan, *speculatores* are missing. Trajan seems to have replaced them with the *hastiliarii*, an escort chosen from the new *singulares*-bodyguard.[40]

Domitian, founder of the equites singulares Augusti?

During the siege of Jerusalem in 70, Titus had a guard of 600 or more picked horsemen, including a number of archers. Chosen from the cavalry *alae* in the eastern armies, these *equites singulares* stood in a tradition of horse guards for provincial army commanders that reached back to Augustus. Now, however, they served an emperor. For his personal safety and as an ever-ready crack force they went wherever Titus went, reconnoitering, saving the tenth legion and its siege engines from enemy sallies, entering the breach of the city wall, and beating off a last attack by the besieged. After the war, his soldiers – surely his guardsmen – begged Titus, some even with threats, to take them along to Rome; yet he balked in the end, for to do so, and to come with an armed elite guard to the city, would look as if he wanted to overthrow his father. Instead, Titus sent his guardsmen to Syria as a regular cavalry unit under the name of *ala I Flavia praetoria singularium*.

Though they were not taken to Rome, Titus' *equites singulares* had shown that it was possible to draw an outstanding horse guard from regular auxiliary cavalry – indeed from Eastern horsemen and archers. Domitian, it seems, took heed.[41]

In 83 Domitian fought the Chatti and, like all emperors in the field, needed a horse guard. By then the horsemen of the *alae* who under Vitellius in 69 had become praetorians were no longer young. The emperor thus may have relied on regular praetorian horsemen or he may have raised a guard of *equites singulares*, as his brother Titus had done. Domitian also needed an outstanding horse guard in 85 and 86, when he went to the Danube for the Dacian wars, and again in 89, when he returned to Mainz to deal with the revolt of Saturninus. But is anything known of such a guard?

It so happens that the oldest known gravestone of an *eques singularis Augusti* was found in Mainz (Plate 1). Set up for Flavius Proclus from Philadelphia (Amman/Jordan), the striking relief shows an Arab horseman. Wielding a composite bow made of wood and bone, he pulls the bowstring with middle and index fingers in the 'western release'. The stone can be dated on stylistic grounds to the end of first century. Perhaps, then, Proclus belonged to Domitian's guard when the emperor stayed in Mainz during the winter and spring of 89. To be sure, Proclus could also have come there as a letter bearer, or he might have belonged to Trajan's horse guard when that emperor passed through Mainz in 98 on his return from Cologne. So elaborate a gravestone, however, points to a longer stay at Mainz, which makes it more likely that Proclus served in Domitian's guard.

If in 89 Domitian indeed had a horse guard of *equites singulares Augusti*, he must have brought it with him from Rome, as Caligula brought his *Batavi* in 39 to Mainz to quell a revolt of the garrison there. To his Danubian wars in 85–6 Domitian had summoned a corps of eastern bowmen, some of them from Philadelphia like the guardsman buried in Mainz. Following in his brother Titus' footsteps, Domitian could have chosen some of these Easterners to become his *singulares* horse guard and may have taken them with him to Rome. Although they would have been there in 87–8, and again in 89–92 and 93–6, no gravestones or altars of the horse guard have come to light from those years. We cannot say for certain, then, whether it was Domitian or Trajan who took the great step of bringing a horse guard of frontier soldiers to the city. If it was Domitian, the famous cavalry battle in the circus in 89, entertaining the masses for the greater glory of the emperor, was fought by *equites singulares Augusti* trained by an Easterner, a 'Greek nobody'.[42]

Among Domitian's murderers there was a trainer of gladiators, perhaps a drillmaster in his guard. But whatever guard Domitian had, they did not save him. Since emperors kept a wary eye out for how their forerunners had been murdered, Domitian's death led Trajan to look for a horse guard that was as faithful as it was skilled – one, in short, like the Julio-Claudian *Germani corporis custodes* whom Suetonius called:

> The German cohort, raised by the emperors as a bodyguard and in countless cases found altogether faithful.

Trajan, therefore, raised his own horse guard, again in Lower Germany.[43]

2
RIDING HIGH:
THE SECOND CENTURY

*Hadrian lived most of the time in peace with foreign nations, for
they saw that he was battle-ready. So well had his army been
trained that the so-called* Batavi *horsemen swam the Danube
with their weapons. When the barbarians saw this, they became
afraid of the Romans.*

Dio 69,9.

For the second century we have outstandingly rich sources. A wealth
of inscriptions, triumphal reliefs, gravestones, and altars, bring the
horse guard to life. To be sure, archaeological finds furnish facts rather
than insights, and images rather than events, yet several literary
sources fill in the picture.

Raised on the Rhine

In late 97 the praetorians rioted, demanding that Nerva avenge
Domitian's death. The ageing emperor had to give in, and the rioters
killed several men dear to him. His authority was badly shattered. To
save what he could, Nerva adopted Trajan. As governor in Germany,
Trajan commanded a large army, strategically well-placed to back
him up and give pause to unruly troops in Rome.

Among the household troops in Cologne, Trajan had the governor's horse guard, some five hundred hand-picked *equites singulares
consularis*. Chosen from the *alae* in the province, they were Batavians,
Ubians, and men of the other tribes that had contributed the
Julio-Claudian *Germani corporis custodes*. Trajan, as Caesar, may have
made them his own guard and rounded out their number to one
thousand, the strength he preferred for elite cavalry units.

In January 98, when Nerva died, Trajan became Augustus and his guardsmen became *equites singulares Augusti*. On military diplomas they are called 'horsemen of our lord' (*equites domini nostri*) which shows that *singularis Augusti* must be understood as 'the emperor's own'. The other, more common meaning of *singularis*, 'matchless' or 'outstanding' was perhaps also understood all along and welcomed by troopers and emperors alike.

The new guard also came to be called *Batavi*, like the *Germani corporis custodes* of old. *Batavi* as a name for the horse guard was still in use a hundred years later, under the Severan emperors, and fittingly, among the *equites singulares Augusti* in Rome the traditions of Lower Germany outweighed those of all other nations.[44]

From Caesar to Galba, for over 125 years, the *Germani corporis custodes* had served as the emperors' horse guard. They had been the last to abandon Nero, and history remembered them well as skilled and faithful horsemen. Now in 98, less than thirty years after their fall, Trajan raised his guard again from the same tribes and with the same tasks of serving as a bodyguard and as a crack fighting unit. By giving back to the Batavians and their neighbours the ancient honor of furnishing the emperor's bodyguard, Trajan gained for himself the finest and most faithful horsemen as well as the loyalty of Lower Germany and of the Batavian *auxilia* at a time when his grip on power was still new. The Batavians had been reviled under the Flavian emperors, but Trajan now refounded the town of Ulpia Noviomagus at Nijmegen as the new Batavian capital, the hometown also of many future guardsmen.[45]

A comparison of the homelands of Claudius' and Nero's German horsemen with those of the second century (Fig. 2), shows that the Batavians and the Ubians (or their towns of Ulpia Noviomagus and Claudia Ara) provided the lion's share of guardsmen in both periods. Remarkably, the seaboard tribes are missing from Claudius' and Nero's guard, even though Nero, when still a prince, had a Frisiavo in his guard. The reason for their absence may be that when Caligula came to the lower Rhine to make war and recruit horse guardsmen, the Canninefates stood aloof. Tungrian guardsmen are missing in both periods which suggests that in the second century the tribe belonged to Gallia Belgica rather than to Lower Germany, or that the Tungrians were not seen as Germans.[46]

In his *Germania*, Tacitus says German leaders were wont to have a band of picked young warriors around them (a *comitatus*) for splendor in peace and protection in war, men whose oath (*sacramentum*) was to defend and safeguard their leader (*princeps*), and who readily ascribed their own brave deeds to him. Using terms of the emperor's court, the

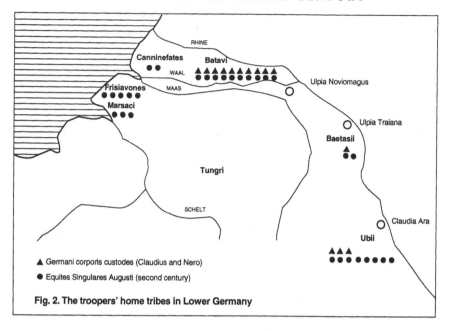

Fig. 2. The troopers' home tribes in Lower Germany

oath of the emperor's guardsmen, and the ethics of Roman officers, Tacitus may have had the *Germani corporis custodes* in mind, the *comitatus* he knew best. His *Germania* was published in 98 when Trajan spent the better part of the year on the Rhine, facing Germany and dealing with the Batavians. Tacitus thus seems to have highlighted the *comitatus* and the *virtus* of the Batavians – the manliest Germans West of the Rhine – because Trajan was restoring the *Batavi* horse guard. The fame of the former *Batavi* guard was untarnished, and when Suetonius called the *Germani corporis custodes* 'an altogether faithful guard' (*cohors fidelissima*), he owned that it was the trustworthiness of these tribesmen as much as their fighting skill that made the emperors look to them for a horse guard.

The Batavians, too, looked back proudly on their service as bodyguard of the Julio-Claudian emperors, for they passed on to their warrior sons the names, some of them Greek, they had borne as guardsmen in Rome. Caesar's praise for the *fides* of the German horse guard in 46 BC shows that it was not something the *Batavi* had learned in Rome. It was a cultural trait they had brought with them. Since Tacitus is right about German faithfulness, he is likely to be right about the *comitatus* as well. If so, Caesar as well as Trajan, in raising bodyguards, not only hired German horsemen, but adopted a German institution.

40

From Cologne, Trajan went to Pannonia, wrapping up the war against the Marcomanni and Quadi, and then on to Moesia, to shore up the armies facing Dacia across the Danube. Perhaps he also chose some horsemen from the Danubian armies for his guard, for men from the Danube later served in the *equites singulares Augusti*, and Trajan may have picked them to honor their armies and to win their fighting skills for the guard. Yet the horse guard's Lower German tradition throughout the second century was so strong that it must have prevailed at the outset. It stands out in the choice of gods they worshipped and in their gravestone reliefs. The supper in the gable of a gravestone and, beneath the inscription, the groom, long-reining the horse with a coil of reins in his right hand (Plate 2), are together typical Lower German motifs. Trajan, therefore, raised most of his *equites singulares Augusti* in 98 in Lower Germany, even though they cannot be shown by documentary sources to have been fully constituted under a tribune before 110. In late 99, Trajan came to Rome and thrilled – or frightened – the citizens with the sight of a thousand tall, blond, and blue-eyed horsemen, the city's new horse guard. They stirred the poet Martial into writing verse.[47]

Entry into Rome

Horsemen were Trajan's main escort as he came up the Flaminian Way. Martial had sighed in anticipation of the happy day when Rome would behold her emperor coming from the north:

> Happy those whom fate gave to see
> the emperor glowing with northern suns and stars!
> When will there be the day when field and tree and every
> window will shine, graced with a Latin lady?
> When will there be the sweet delays, the long dust trail
>   behind Caesar,
> and all Rome to be seen on the Flaminian Way?
> When will you, horsemen, and you, Moors in embroidered
>   Nileland tunic
> draw near, and all people shout together 'He is coming!'?

Martial's 'horsemen' are guards rather than knights, for among aristocrats, senators would have taken pride of place over knights. Besides, only bodyguards had their place next to Mauri outriders (*cursores*) in the emperor's following, and indeed, Mauri belonged to the stock image of the horse guard: a North African is shown on the arch of Galerius, and a curly-haired horseman shown riding with the emperor on a scene of Trajan's Column, may be a Maurus as well. Martial's poem about the emperor's return to Rome thus balances the relief of his setting out (see Plate 4). In the imperial ceremonies of *profectio* and *adventus*, the horse guard played an outstanding role.

At the sacred city limit, Trajan alighted from his white stallion and made his way on foot. The horse guard, and even the lictors, fell back while Trajan mingled, openly and unguarded, with the senators and the foremost knights, a thing dear to the heart of a man like Pliny. Indeed, one of the highest honors an emperor could bestow on a group of fellow aristocrats was to meet them without his bodyguard – Constantine thus honored the bishops at the council of Nicaea in 325.

'The soldiers looked no different from the people and were as mild and meek!' Pliny gushed. Behind his wishful thinking and flattery there lurked the hopes and fears of Rome's leading men as they saw the emperor thronged by battle-hardened soldiers. An emperor's entry into the city could be far grimmer, as Severus and his frontier troops were to show in 193, and our sources all agree that Trajan, though a warlike emperor, firmly kept his soldiers from getting high-handed and overbearing, a matter of great concern not only in Rome but wherever the emperor traveled, for his guard was billeted in the towns along the way. Indeed, emperors were widely judged by how devastating their visits were – beginning with Cicero's letter about Caesar and his guard mentioned above.

In Roman eyes, Trajan, the 'best of emperors', may have set things right by bringing back the *Batavi* horse guard, for it was part of the imperial constitution, hallowed by Augustan precedent. In the provinces the institution of a double horse guard had survived: governors kept *equites singulares* as well as *equites legionis* about themselves. Trajan restored this twin aspect to the emperor's guard. On his Column auxiliaries do most of the fighting, while legions and praetorians back them up. With *equites singulares* as well as *equites praetoriani*, the emperor's horse guard, too, could now strive after this ideal.

A further cherished tradition was to have the horse guard trained by battle-seasoned, highly decorated centurions. Domitian had used a 'little Greek' (*Graeculus magister*) instead. Now Trajan restored the old custom and made four high-ranking centurions training officers (*exercitatores*) of the horse guard and, through them, of all Roman cavalry.[48]

Singulares and praetorians

Once in Rome, Trajan did not lodge his *singulares* in the praetorian fortress, though room might have been found there (in the third century it housed several thousand soldiers more than in the first). Batavians, however, were known for brawling with other troops, which was reason enough to give the new horse guard, like the earlier *Germani corporis custodes*, a fort of their own. Built on the Caelian Hill, away from the praetorians, it was also much nearer the palace than the former fort of the *Germani* across the Tiber. When the horse guard came back from the first Dacian War in 103, their fort was ready, if one may judge from the short time it took to build the next fort under Septimius Severus. They established a graveyard of their own and, unlike the *Germani*, had their gravestones carved in the Lower German manner.

The coming of the new horse guard was a blow to the praetorians, for if they now fell out with the emperor, he, unlike Nerva, would have another military force to turn to. Worse, the praetorians now had to share not only the power that came from serving the emperor, but the glory too. It seems that the task of escorting the emperor, hitherto the privilege of the praetorian *speculatores*, under Trajan went to the *hastiliarii* of the horse guard. Though no source states this directly, the fact that the *speculatores* are no longer heard of as emperors' escort after the year 69, and that the *hastiliarii* are known in that role by 135, suggests that the change-over happened in 98. If so, it was brought about by the praetorian mutiny against Nerva.

One of Trajan's first steps after he became emperor had been to send for the praetorian prefect and his accomplices. Trajan gave out he was to employ them with him in Germany. Instead, he had them killed. The praetorian mutineers must have been a specific group, so specific that the emperor could claim, without arousing suspicion, he needed them in Germany – clearly they were the *speculatores*. They had betrayed Galba and may likewise have bullied Nerva. Having killed a number of praetorians, Trajan could not trust soldiers of that unit with his life. Instead, he kept his *singulares*-guard around him: they had escorted him as governor and now they escorted him as emperor. A hundred years later, in a similar case, Septimius Severus did the same. New rulers tend to need new guards, hence guard units throughout history were raised and disbanded faster than units of the line.[49]

After ruling Rome for some 150 years, the emperors felt little need to hide the 'foreign' soldiers who underpinned their power, all the more since the *auxilia* by this time were far along the road to romanization and no longer truly foreigners. The *hastiliarii* of the bodyguard could therefore well serve as Trajan's escort. If the *Germani corporis custodes* once had lacked some of the trappings of true soldiers, the new *equites singulares Augusti* were soldiers in every way. They had fully Roman names and, very likely, citizenship. Their ranks included the essential cavalry under-officers of decurion, *duplicarius* and *sesquiplicarius*, and their commander was a tribune of the guard. The unit was called a *numerus* not because it was unroman, but because that was the traditional title for the *Batavi* as well as for the units of provincial *equites singulares* from which Trajan had raised his guard.[50]

Trajan's name for the unit fits this mold. Augustus had stressed the civilian side of his reign with the title *Augustus*, while the more military term *imperator* was the popular title for the emperor from Trajan's time onward. Trajan, however, named his guard *equites singulares Augusti*, not *equites singulares Imperatoris*. He may have followed a tradition, but he may also have wished not to ruffle the feathers of dreamers who hankered after the old days when the senate ruled and the city was free of soldiers.

The horse guard's name thereafter mirrors the emperors' ever-growing high-handedness. Soon the title *imperator* replaced *Augustus*, and under Commodus the title *dominus noster*, 'our lord', in turn replaced *imperator*, heralding the absolute monarchy. Minor variations aside, the name of the unit thus changed as follows:[51]

first century	*Germani Corporis custodes*
second century	*equites singulares Augusti*
later second century	*equites singulares imperatoris nostri*
third century	*equites singulares domini nostri*

In the field with Trajan

Trajan's Column in Rome, commemorating the emperor's warlike *virtus* in the Dacian wars of 101–6, portrays the horse guard no less than seven times: escorting the emperor on the march, coming to the rescue of embattled troops, traveling down the Danube aboard ship, reconnoitering, and on parade. The horse guard, therefore, more than any other unit, shared Trajan's vaunted battlefield deeds; its pluck and skill must have pleased the emperor. The guardsmen, one would like to think, were the 'many soldiers' Trajan knew by name or nickname and deed.[52]

In 105 the horse guard rode with the emperor from Rome to the theater of war. A scene on Trajan's Column shows the emperor and two officers in long-sleeved tunics riding in front. Behind them a third officer, wearing a mailshirt and looking back towards the soldiers, may be the tribune of the horse guard. The troopers in the background wear mailshirts and the light cape (*sagum*). Those in the foreground, however, wear cloaks held together over the chest rather than over the right shoulder, and they lack armor. If indeed two groups are distinguished here, they might be the *hastiliarii*-escort in the foreground and regular *equites singulares Augusti* in the back. Since danger is still far away, no helmets are worn, and shields are slung to the sides of the horses. The scene bespeaks the horse guard's essential duty of escorting emperors on their travels through the provinces.

In another scene, Trajan and his horsemen come to the rescue of a beleaguered army. Now they are battle-ready. They have taken off their capes, donned their helmets and gripped their shields with their left hands. The spears they once wielded in their right hands are now gone. Fine, wavy lines make it clear that their shirts, with zigzag hemlines, are made of mail, not leather. Their shield emblems differ from each other and from the one shown on a gravestone (see Plate 8), which suggests that the shield emblems on the Column were invented by the artist. The emperor, as usual without shield and helmet though wearing a cuirass, greets the beleaguered army with the horseman's salute, an open right hand. He will then have sent the guard into battle, while he himself reined in and stayed to the rear. The tribune of the guard is not shown – another leader of the horsemen would have cramped the emperor's presence.[53]

Guarding the emperor could be dangerous, for to bolster the morale of his men and to witness their brave deeds, a ruler had to share some of their risks. During the siege of Hatra (Al Hadr in northern Iraq) in the Parthian war, Trajan left the safety of the siege works and rode up to the city walls as he sent 'the cavalry' – surely his own guardsmen – to storm the wall. He had taken off his purple coat so as not to be recognized. But when the gunners on the wall 'saw his proud grey head and his awesome looks, they guessed who he was and shot at him, killing a horseman of his escort'.

Since it envisions camel riders as battle troops, Hyginus' treatise on the layout of an imperial field-army camp may have been written for Trajan's Parthian war. As to the horse guard, this may well be so, for in Hyginus' camp, the *equites singulares Augusti*, up to 900 strong, had as prominent a place as they had in the war itself.[54]

Swimming the Danube under Hadrian's eyes

The most famous poem about a Roman horseman – perhaps written by Hadrian himself – praises a soldier who, in the summer of 118 with all his weapons and under the emperor's eyes, swam across the Danube:[55]

> I am the man, once well known to the river banks in Pannonia,
> brave and foremost among a thousand *Batavi*,
> who, with Hadrian as judge, could swim the wide waters
> of the deep Danube in full battle kit.
> From my bow I shot an arrow which, while it hung in the air
> and fell back, I hit and broke with another.
> Whom no Roman or foreigner ever outdid,
> no soldier with the spear, no Parthian with the bow,
> here I lie, on this ever-mindful stone have I bequeathed my deeds
> to memory.
> Let anyone see if after me he can match my deeds.
> I set my own standard, being the first to bring off such feats.

The one thousand *Batavi* of the poem seem to be Hadrian's horse guard of the *equites singulares Augusti*, returning from the Parthian war, for *Batavi* was one of the guard's names, and the historian Cassius Dio, describing the same feat, says that the swimmers were horsemen. The feat astounded the tribes of the neighborhood, and was thus not a routine exercise as it would have been if performed by a local cavalry regiment. As shock troops, all horse guards, German, Roman, or Parthian, excelled in contested river crossings. Once the shallows of the far bank were reached, the fight might be first against enemy archers, hence the emperors' horse guard needed to take its own bowmen across in the first line, bowmen such as the Syrian Soranus honored by the poem.

To the tribes beyond as well as to the Romans, broad rivers were trusted defense lines. Dio therefore adds that when the nations from north of the Danube saw this feat of the *Batavi*, they gave up all thought of war against Rome and even asked Hadrian to settle their quarrels. The horsemen, combining German boldness with Roman discipline and Eastern skill in archery, here, as with Trajan before Hatra, were the cutting edge of the imperial field army and as such of strategic importance. To be sure, set against the Roman army's overall strength of nearly 400,000 men, 1000 horsemen may seem too few to carry much weight. But actual field armies often numbered only a few thousand men, and 1000 horse, highly trained, could outmaneuver and overwhelm much larger numbers of the enemy, witness not only Dio's stated testimony but also the example of the Marcomannic bodyguard cited above. Likewise in 364, according to Ammianus Marcellinus, fewer than 500 Germans from the lower Rhine area swam the swollen Tigris, and this so struck the Persian Shah that he sent offers of peace.

Of German rulers Tacitus says: 'Not only in his own nation but in neighbouring states, too, he has a name, he has glory, who has the most and the fiercest guardsmen. Such leaders are sought out by embassies, are given gifts, and often curb wars by their fame alone.' If this was indeed so, then Hadrian's own *Batavi* horsemen may have suggested swimming the Danube as a show of force, and it worked because the nations north of the Danube were likewise Germans.[56]

A plot against Hadrian's life thwarted

During the first year of his reign, in 117, Hadrian shocked the senatorial aristocracy by killing four leading ex-consuls on charges of high treason. The charges are widely thought to have been trumped-up, merely to get rid of potential enemies. But they ring true in the light of Calventius Viator's career. Viator was a centurion and drill master, and very likely also commander, of the governor's horse guard in the Dacian capital of Sarmizegethusa. There he set up the following altar:

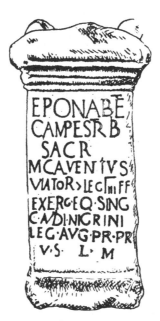

Fig. 3 Viator's altar from Sarmizegethusa.

47

To the goddesses of the horses and
the goddesses of the training field
sacred.
M. Calventius
Viator, centurion of the fourth legion Flavia felix,
training officer of the horse guard
of C. Avidius Nigrinus,
the governor.
He gladly and deservedly fulfilled his vow.

Epona (or the Eponae) and the Campestres were worshipped together
by the horse guard in Rome, since they stood for care and skill in
horsemanship. Calventius Viator thus may have seen service with the
equites singulares Augusti, perhaps as a decurion, the rank from which
one rose to legionary centurion. Skill in the exercises of the emperor's
horse guard was the best qualification for becoming training officer in
a governor's horse guard. Besides, Viator may also have been sent to
Dacia to keep imperial headquarters informed of the goings-on in the
province.[57]

There was indeed much going on. If the *Augustan History* is to be
believed, the governor Nigrinus meant to murder Hadrian at a
sacrifice. The emperor escaped the plot and in its aftermath had
Nigrinus killed, together with three other ex-consuls, his fellow-
conspirators. Surprisingly though, Viator, who had commanded the
guard of the faithless governor, afterwards became lieutenant-
commander of Hadrian's own horse guard. It looks as if Viator was
promoted for having betrayed Nigrinus' plot and saved Hadrian's life.
Hadrian no doubt heard of the foul play when he was in the
neighbouring province of Lower Moesia, for there he announced it,
and there also was stationed legion V Macedonica whom Viator now
joined as a centurion, presumably of very high rank.

Prior service in the horse guard would explain how and why Viator
could contact Hadrian to warn him – as an officer in Trajan's guard he
would not only have known Hadrian personally but could have kept
in touch with him. Snooping was an established task of headquarters
troops: Hadrian sent soldiers as spies everywhere to keep himself
informed about senators and would-be rivals. Promotions of

5 *A guardsman of Septimius Severus during the emperor's victory speech in
Ctesiphon: from Severus' arch on the Forum in Rome. Some horsemen still took the
field when they were 50 years old.*

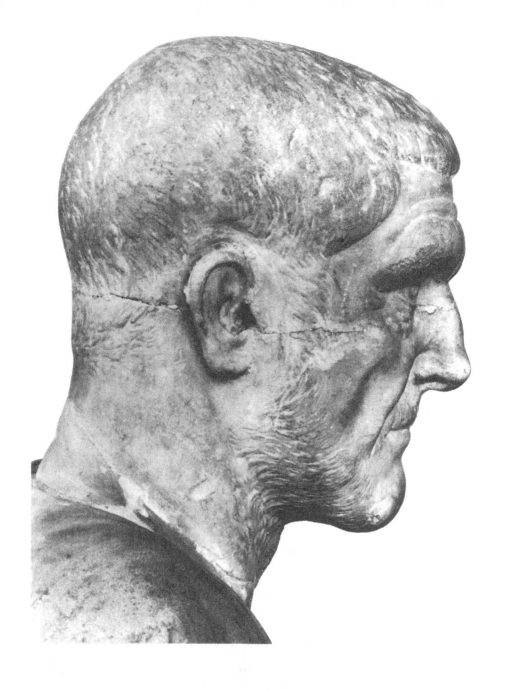

guardsmen to legionary centurionates thus not only ensured uniform training throughout the army, but also put men whom Hadrian trusted into key positions in provincial armies.[58]

Hadrian's actions belie charges about a trumped-up plot, for he entrusted Viator with his life by making him commander of his escort. In 128, Hadrian's famous speech on the African legion's parade ground in Lambaesis ends with words of some intimacy for the man who had saved his life: 'Now, Viator, for the Commageni camp!' And as late as 130, thirteen years after the foiled plot, Viator was still in charge (*curam agens*) of eight squadrons, one fourth of Hadrian's horse guard, then wintering in the Arabian town of Gerasa during the emperor's stay in the Orient. Viator's reward was such that the plot must have been real rather than fictional.

As a legionary centurion and *curam agens*, however, Viator was not commander of the whole horse guard. True commanders had to be tribunes, and a tribune of the horse guard is known in 120, during the years 117–30 when Viator commanded Hadrian's escort. Viator thus ranked below commander. Perhaps he was in charge of the *hastiliarii* or of an *ad hoc* escort of several squadrons. Most likely he kept the title *centurio exercitator* that he had held under Nigrinus, unless one is to call him *a pugione* as commander of the escort. The reason why he never became tribune must have been his lowly birth – class distinctions were still strictly upheld during the second century.

Hadrian had renewed the exercise regulations for cavalry, and, no doubt, he had worked them out with the *centuriones exercitatores* of the horse guard. Perhaps, then, during his maneuver critique at Lambaesis, Hadrian had Viator along not only for his safety but also as an expert adviser in cavalry training.

No emperor traveled as much as Hadrian, and wherever he went the guardsmen followed him. To serve as his bodyguard meant long years spent criss-crossing the empire. No wonder that few of the second-century *equites singulares Augusti* were married. Troopers hated to travel under arms, yet this was the guard's task. Since they set up gravestones only where they stayed for some length of time, but one headstone bears witness to Hadrian's wanderings. It comes from Heraclea Lyncestis (Bitolj) in Macedonia on the *via Egnatia*, the trunk road to the East (see Fig. 9) where the guard may have stayed for the winter, perhaps as an advance party for Hadrian.[59]

*6 Emperor Maximinus, once a horse guard trooper. His short-cropped soldier's haircut and strong-willed frown stress achievement won by toil (*labor*).*

Marcus Aurelius' guard

Antoninus Pius, who never left Italy throughout his long reign, struck off the rolls many of Hadrian's older guardsmen as soon as he came to power. His peaceful reign must have been a quiet time for the horse guard, spent mostly in ceremonies or at the circus, when not in weapons training or on guard duty. Still, at least one guardsman is known, whom Antoninus Pius placed as an officer in a frontier unit to ensure the troops' training and loyalty: a certain M. Cocceius Firmus, centurion of the British second Augustan legion. Though he does not say so, the altars Firmus set up at the Antonine Wall prove him to have been a former guardsman, and he may have gone to Britain to watch out for the emperor's interests, as Viator had done in Dacia for Hadrian.

The horse guard's last duty to Antoninus Pius was also their first to Marcus Aurelius and Lucius Verus. On reliefs at the base of the Antonine Column, guardsmen and knights are shown riding in Pius' funeral parade. Their sequence is not wholly haphazard or merely artistic but, in the context of the pomp and ceremony of the capital, betokens the horse guard's duty of watching over the emperor's safety: three horsemen ride before and three behind the new ruler. In order to add lightness and elegance (*gratia*) to the show, they wear no mailshirts.

Under Marcus, war engulfed Rome and the guard. Written accounts of the Marcomannic wars are largely lost, and triumphal reliefs are our best source, though they hint only dimly at the horse guard's deeds. One of the Aurelian panels of a lost victory monument sets forth in detail how the emperor in a grand ceremony joined the horse guard as he left Rome for the front (see Plate 4). A medallion of 178 likewise shows Marcus and Commodus riding off to war (*profectio*): the two emperors on their steeds, wearing cuirasses and holding staffs of command, are portrayed as warriors on horseback.[60]

The Aurelian Column shows the horse guard several times, most strikingly in a parade scene unparalleled in Roman art (Plate 3). Four soldiers and their horses stand at attention, the men smartly aligned to the left of their steeds and each in the same parade stance (compare Plate 7) – right foot forward, eyes sharply right, left hand on the spear, right hand on the bridle, holding the head of the horse towards the trooper. All four steeds are stallions. Their saddles end in the usual zigzag hem and are set over fringed blankets. The men, in the ring-topped helmets commonly found on the two Columns, wear long, fringed all-weather greatcoats, like the cloaks of the *speculatores* held together over the chest rather than clasped over the shoulder. Their spears show no blades, either because they were pointed downwards or because the men were the emperor's *hastiliarii*-escort who had replaced the praetorian *speculatores*, and who as military police used the shaft (*hastile*) rather than the blade of their weapons.

Marcus Aurelius in this scene rides along the row of horsemen to review them. The rider before him, armed like a common soldier, is perhaps a training officer (*centurio exercitator*). The officer behind the emperor, wearing a long-sleeved tunic, a cape, and a sword, but no cuirass, is the tribune of the horse guard, judging from his bell-shaped officer's helmet. No other picture in Roman art and no literary report tells of a unit standing at attention while an officer rides along its front. The scene nevertheless shows that leaders reviewed soldiers on parade much as they do today.

Another striking scene from the Aurelian Column shows the army advancing into enemy land north of the Danube. Marcus, thronged by his *equites singulares*, bears no weapons at all, not even a cuirass, while his guard is fully armed with body-armour, helmet, and sword. They are holding spears and shields at the ready. Their armor alternates between mail and scale so schematically that the reason must have been purely artistic, hence little can be learned here about the cuirasses actually worn by the guard. Unlike Trajan, however, the less warlike Marcus rides not at the head of his guard but in their midst. Both scenes thus skillfully stress the emperor's safety amid his guardsmen but also betray his greater dependence on them.

When they died in the field, guardsmen, like other troops, were given gravestones only if their unit stayed for some time at the place of their death. This was the case with Marcus' stay at Carnuntum, the emperor's field headquarters in Pannonia during the Marcomannic wars of 171–80, where the headstone of a guardsman, though broken, has come to light.[61]

Commodus, the soldiers' ward

When in 180, at the beginning of his reign, Commodus came back to Rome from the Marcomannic war, the official city bulletin (*acta urbis*) proclaimed: 'The gods have given the emperor to the army and the senate to safeguard forever in the Commodian palace'. It was the army's duty to guard the emperor, for in imperial ideology the welfare of the empire hinged on that of the emperor. Tellingly, the *acta urbis* name the army first, before the senate. The frontier armies and the guard were to fend off any need for Commodus to take to the field, for after Hadrian's travel-filled reign and Marcus' endless wars, the emperor's staying in Rome was now seen as a sign of strength.[62]

In 182, the bodyguards had a chance to show what they could do. The young senator Quintianus, a former friend and companion of Commodus, joined a plot against the emperor. In the dark, narrow entryway to tne hunting theater, he pulled out a dagger, ran towards the emperor, and yelled 'See, this is what the senate sent you!' Those were too many words. Before he could strike the guards had laid hold of him. This is what the troopers had been hired for and the outcome must have buoyed them in their own eyes as well as in those of the emperor.

Commodus, grown wary, now raised the number of bodyguards on duty. What is more, by 186 at the latest, he bestowed the command of his bodyguard on lowly Cleander, a former Phrygian slave, whose title, unheard of before, was *a pugione*, 'Wielder of the Emperor's Dagger'. Perhaps Cleander headed only the *hastiliarii*, the imperial escort, but he may also have replaced the tribune as the horse guard's overall commander – not so astonishing for one who little more than a year later rose to the rank of praetorian prefect. The strengthening of the bodyguard and Cleander's new rank *a pugione*, noted by inscriptions and historians, betray a shift in the tasks of the horse guard: during Commmodus' reign it served less as a strategic horse guard and more as a bodyguard.

Cleander's rise to praetorian prefect parallels the careers of other commanders of the *equites singulares Augusti* who likewise reached that rank. Having headed the horse guard, then, was a mark of high distinction. Cleander's career, however, ended unlike that of other prefects, for he was sacrificed to a rioting mob. Like Spiculus, the commander of Nero's bodyguard, Cleander had drawn fierce hatred upon himself, so much so, that two hundred years later 'more hateful than even Cleander' were still winged words. Some of that hatred may have been aimed at the horsemen, too.[63]

Like the other 'bad' emperors, Commodus slid back into the autocrat's way of choosing freedmen rather than army officers for the command of the horse guard. Like them, having fallen out with the senate, he turned more and more to the guard for backing. By switching his residence from the Palatine hill to the Caelius in order to be nearer the gladiators, he also moved, by coincidence or design, nearer to the fort of the *equites singulares Augusti*. To win over the soldiers, he slackened the reins of discipline. He allowed the praetorians (and surely also the horsemen) to beat people with nightsticks – nasty weapons of riot control – and even to carry axes for breaking into houses to loot them. This is the stock-portrait of a tyrant, painted by Dio who, as a senator, hated Commodus, and it has been highlighted with even more garish colors by Herodian. It may nevertheless be true, as the events after Commodus' death showed when the praetorians insisted on honors for their former emperor. Nor is there evidence that this 'offended the provincial armies and aroused their jealousy' as has been said. On the contrary, Septimius Severus, so beholden to the Danubian army, restored Commodus' good name – no doubt in line with the wishes of his soldiers.

The bond between the emperor and the guard was at work also in the personal answer Commodus gave in 182 to the petition of a certain Lurius Lucullus. The emperor's letter to Lurius promised help against wrongs done to the small leaseholders on the African estate of the Saltus Burunitanus, the home, it seems, of Lurius. The historian M.I. Rostovtzeff went so far as to say: 'I am fairly confident that Lurius Lucullus was a soldier, probably one of the soldiers stationed in Rome, not a praetorian (as he was of provincial origin) but an *eques singularis* or perhaps a *frumentarius*.' His guess turned out to be true, for a newly-found gravestone of a horse guardsman cites a troop leader named Lurius, and since that name is not otherwise found among the horse guard, he seems to be the petitioner or a relative of his. Commodus himself answered the petition since as a decurion of his guard, Lurius was very much his man. If Lurius had served him well, Commodus was bound to him by obligation (*fides*). The emperor heeding a former under-officer of his horse guard, now leaseholder on an imperial estate in Africa, betokens the unfolding of the empire's new military aristocracy. Wherever former soldiers of the guard settled down, they became local leaders since they could gain a hearing.[64]

Commodus hoped to heighten the magic of being emperor by passing for the greatest gladiator and hunter of wild animals in the arena. For teachers he took the finest Parthian bowmen and Moorish spearmen. Since there were Parthian as well as Moorish marksmen in the horse guard, and since emperors were wont to train with their guard, Commodus' teachers no doubt were *equites singulares* – there is no need to think of a separate unit of Moors and Parthians stationed at the Castra Peregrina in Rome. Commodus strove hard to outdo his teachers. Unlike Nero, who had animals in the circus slaughtered by the horse guard, Commodus did the slaughtering himself. On one occasion he speared and killed 100 lions with exactly 100 spears, always finishing the animals at the first throw. As a bowman, he shot off the heads of ostriches, using half-moon shaped arrowheads: the birds ran at full tilt even after loosing their heads. All this did not endear him to the upper classes, but the soldiers liked him well.[65]

Since Commodus shied away from military exertion, the praetorians lost their mettle, growing soft in the easy life of the city. Senators, who knew the frontier armies, were overcome with laughter at the sight of the praetorians exercising, and when in 193 Septimius Severus drew near with his frontier troops, the guard did not dare to fight, quite unlike the gallant stand Otho's praetorians made in 69. The horse guard, though, may have fared better, for it renewed itself from experienced warriors who, as a rule, had served for five years in the frontier *alae* before they joined the guard. They thus kept up not only their fighting skills but also their connection with the frontier armies, while the praetorians, recruited in the inner provinces, became isolated in their scorn of the 'barbarian legions' or the 'less trustworthy provincial branches of service'. It was not by luck alone, therefore, that the *equites singulares Augusti* weathered the crisis of 193, while the praetorians went under.

With Commodus' death, the days were gone when an emperor and his guard could tarry in Rome for years on end, enjoying life in the city without being put to the test of weapons. The thought of having it this good, though, lived on in wishful thinking: more than a hundred years later Maxentius is said to have boasted to his guard that they ruled together and lived it up, while others stood guard on the frontiers for them.[66]

Pertinax' murder

When in 193 a murderous mob of praetorians forced its way into the palace, Pertinax personally went to face them and tried to reason with them. He need not have done so. He could have killed them, we are told, with the help of the nightwatch and the horsemen at hand. Once more we find a sizeable horse guard detail on duty at the palace, as in Caligula's time.

The mutineers, after some wavering, killed Pertinax. The first to strike the emperor was a certain Tausius of the Tungrians in Gallia Belgica or Lower Germany (see Fig.2). Since many horse guardsmen but few praetorians came from Lower Germany, scholars saw in Tausius an *eques singularis Augusti*. However, several gravestones of second-century praetorians from Lower Germany have now been found. What is more, Tungrians are known in the praetorium but not in the horse guard. Dio may be trusted, therefore, when he says the praetorians killed Pertinax. The horse guard, as always, stayed loyal.

54

When Septimius Severus came to Rome with his Pannonian army, he brought his own horse guard and his own praetorians, but while he dismissed the second-century praetorians for having murdered Pertinax, he kept Commodus' horse guard. Certainly he could not have done so had Pertinax' killer been an *eques singularis Augusti*. Its faithfulness, having earned the horse guard a rebirth under Trajan, now allowed it to last through the third century as well.[67]

3

THE ROUGHSHOD THIRD
CENTURY

I captured the emperor Valerian with my own hands.

Shahpur I, Shah of Persia (RG 24)

A dearth of sources

The horse guard of the Severan emperors (193–235) is well known from inscriptions, sculptures, and the remains of their New Fort on the Caelian Hill. Three literary works also deal with this period: Dio, Herodian, and the *Augustan History*. Of these, the eyewitness Dio, who was twice consul, is the primary source and by far the most trustworthy. It is a pity his account survives only in excerpts. Herodian, on the other hand, seeks to come up with melodramatic stories, the *Augustan History* plays recklessly with snippets borrowed from Herodian, and both set the truth at naught. Yet the Severan period is relatively well-documented.

Dio's account ends in 229, Herodian's in 238, and the maelstrom of the mid-third-century crisis swept away most sources for the next half-century. Thus, from 240–90, fifty years of darkness descend, fitfully lit for the horse guard by only five stone monuments. Many emperors fell in battle or were slain at the hands of their own soldiers. Often their guardsmen died with them. Almost everywhere in the empire people, losing heart and lacking confidence in the future, stopped setting up gravestones. At the turn of the fourth century a blaze of news sheds some light, though what it reveals is the fall of the horse guard.[68]

Doubling the horse guard

When Septimius Severus, governor of Upper Pannonia, was hailed emperor at Carnuntum on the Danube on 9 April, 193, he at once made ready to march on Rome. It was a risky undertaking, for his opponent Didius Iulianus sent out armies as well as murderers to foil him. Looking back at the civil war of 69, Tacitus brooded over what use so many regiments were if a single murderer, spurred on by high reward, could strike down a leader – 'it was easier to launch thousands than to escape one'. Severus therefore chose 600 of his best fighters as a bodyguard to be around him day and night. 'They never took off their cuirasses', says Dio, meaning that they never stood down until they reached Rome. During the march, Severus wore cavalry dress and rode on horseback. The 600, therefore, were his horse guard.

Severus, like Trajan, may have raised the bulk of his new horse guard from the *equites singulares* that had served him as a governor's guard. Now they became his *equites singulares Augusti*. There seems to have been something particularly useful about a force of 600 men. Titus rode attacks and reconnoitered with that many guardsmen; and Hyginus, in his treatise on how to build a military camp, likewise reckons with 600 *equites singulares Augusti* in his model camp, even though the horse guard at the time was 1000 strong. A strength of 600 may have been useful in that it was the (full) size of a standard cavalry unit with its battle-tested structure, tactics, and logistics.[69]

On 9 June, two months after his uprising, Septimius Severus and his army paraded through the streets of Rome, a day the horse guard would celebrate for years to come. If we believe the *Augustan History*, Severus' entry was quite unlike Trajan's a hundred years earlier. The whole army, horse and foot, armed to the teeth, followed the emperor. They took up quarters in the palace, porticoes, and temples, as if these were stables. Fear gripped the people as the fighting-men looted the stores and threatened to lay the city waste. If Commodus' praetorians had seemed high-handed and overbearing, Severus' Danubian soldiery outdid them by far. Throwing their weight around, they rioted until they got a rich cash bonus from emperor and senate. Severus showed not much civility either and sent shudders down the spines of the senators by coming to the senate hall with a throng of armed guards.

Severus cashiered the largely Italian-recruited praetorians he found in Rome. Since they had murdered – and not avenged – the emperor Pertinax, they were tainted with treason. In their stead, Severus took legionaries from the Danubian frontier as his praetorians, the men who had proclaimed him emperor. Henceforth the foot and the horse guard hailed from the same, mainly Danubian, provinces. The upright citizens of Rome were shocked at this. Dio says they found the motley guardsmen cluttering the streets 'clod-hoppers to look at, boorish to listen to, and most dreadful to talk to'.

Yet there was good ground for bringing such men to Rome: they were faithful and they knew how to fight. The horse guard had shown this for nearly a hundred years. Ever since Trajan had raised the *equites singulares Augusti*, they had been brought up to strength with men who had served four years or more in the frontier forces. They had become romanized and no longer blindly did the emperor's bidding, as Nero's half-foreign guardsmen had once done. As Romans, however, they acknowledged the same form of loyalty: receiving a favor, a *beneficium*, made a man beholden to the giver and established a formal obligation. Being singled out for service with the guard in Rome was a much sought-after *beneficium* among frontier troops and thus a major reason for keeping faith with the emperor. The horse guard's system of recruiting was now extended to the praetorians as well.

Equites singulares and praetorians, picked from among their peers in the provincial armies, were bound to the emperor for having made them his guards. On the other hand, the emperor was no less beholden to them, for not only did they watch over his safety but they were a link to the frontier armies, some having been chosen for the guard by the rank-and-file of their frontier units. It was wrong to say that guardsmen 'shared the other soldiers' complete lack of political principles and consciousness', for they were faithful, and politics in Rome was about men rather than issues.[70]

The horse guard was not tainted by Pertinax' murder; indeed, they stood ready to fight for him. Hence, unlike the praetorians, they were not cashiered by Septimius Severus. Nor would it have been politic to do so. Dismissing the praetorians had made Italy rife with brigandry, but did little political harm. Sending home the bodyguard, on the other hand, would have made enemies of the auxiliary units along the frontiers from which the men had been chosen, a risk no emperor would lightly take.

Besides, when Severus staged Pertinax' funeral to vaunt his piety and legitimacy, the old horse guard's skill stood him in good stead. Not that the funeral procession involved only military pomp. On view also were the magistrates, the senate, and the knights, as well as bronze statues of the provinces and the guilds, race horses, and, wonderful to hear, 'images of men who had distinguished themselves by some work, some invention, or by their manner of life'. Neverthe-less, in Pertinax' funeral procession the guardsmen shone, all decked out with their weapons, and around the pyre their horsemanship was supreme. The aristocratic knights rode their Troy game, yet they were amateurs while the guardsmen, horseback riders by profession and picked for their prowess, rode warlike attacks, the height of imperial pomp.

Traditionally, the loyalty of the guard was not only to the emperor but to the whole ruling house. This worked well for a dynasty, but it meant that when a new house came to power the emperor had to bring in his own men quickly. The fastest way to do this was to enlarge the existing guard. Severus, it seems, did so by merging the 600 bodyguards that had escorted him from Pannonia with the *equites singulares Augusti* inherited from Commodus.

Having doubled the horse guard, Severus also built them a new fort (*castra nova*) on the Caelian Hill next to what now became the Old Fort (*castra priora*). Both forts were used until the fourth century. An altar to Minerva Augusta from the headquarters building of the New Fort, dated 1 January 197, shows that the guard took up its new quarters upon return from the Eastern campaign of 193–6. No doubt it took a long time to raze the former palace of the Laterani, which occupied the site, and to build the massive vaults underpinning the terrace on which the New Fort rose. Severus, therefore, must have given orders early for the fort to be built, surely during his first stay in Rome as emperor (9 June–9 July, 193). Clearly, he had made up his mind from the outset to double the horse guard. The idea may have occurred to him years before when he saw Commodus' rule upheld so strongly by the guard. Now, as in 69, and again under Trajan, a new dynasty did well in bringing along a new guard.

The Old Fort housed a thousand horsemen, and so did the New Fort. Septimius Severus thus doubled the number of troopers from 1000 to 2000, a telling comment on their usefulness to him. Under Constantine the horse guard (*scholae*) grew even further to a strength of 2500 men. Still, since frontier cavalry regiments (*alae*) were either 500 or 1000 strong, the question may be asked whether the New Fort housed perhaps only 500 horsemen. The tribunes of both forts, however, held the same rank and thus must have led units of the same size. This is clear from inscriptions listing the officers by rank: in 203 Octavius Piso, tribune of the New Fort, was outranked by his colleague in the Old Fort, while in 205 he, in turn, outranked a newly appointed colleague. The tribunes of both forts, therefore, ranked alike, while length of service in the rank determined who came first on the list (the same method was used for ranking cavalry squadrons and their leaders in the frontier armies). Since both tribunes held the same rank, both forts must have housed roughly the same number of men – 1000 each.[71]

How did the troops in the two forts relate to each other? Housed in their own forts, commanded by their own tribunes, and upon discharge receiving their own diplomas, they were units of their own. On the other hand, troops from both forts set up common dedications, worshipped a common Genius, and together made up the *numerus* (corps) of the *equites singulares Augusti*. On their gravestones, therefore, men often did not bother to state in which fort they served. Troopers of the New Fort had their headstones carved in the same style as those in the Old Fort, hence the troops in both forts carried on the traditions of the second-century horse guard.

Inscriptions from the time of the Severan emperors also call the horse guard *Batavi*. If this applied only to the men in the Old Fort they would be set off as *Batavi* against those in the New Fort, who were perhaps Caracalla's *Lions*. More likely, the name *Batavi* applied to troops of either fort. The presence of two units in the same corps led to the highly effective Roman way of raising morale by goading units to outdo one another.[72]

Septimius Severus doubled the horse guard from 1000 to 2000 men in order to discourage uprisings of the sort he himself had undertaken. It was another lurch toward centralizing the empire and turning the emperor into an autocrat, the more so since he also doubled the strength of the foot guard by increasing the praetorians from 10,000 to 15,000 men and bringing the 6000 men of legion II Parthica to Albano near Rome. This, more than anything else, was what senators found wrong with Severus. Placing his trust in the troops' strength rather than in the senators' goodwill, he crowded the city with ruffians and burdened the country with huge outlays. Worse, by taking the field for years on end, Severus set a course that in the long run robbed Rome of her guard and her status as a capital of the empire.[73]

With Severus in the field

Like Caesar's guard, Septimius Severus' *equites singulares* rode and sailed thousands of miles. They fought on three continents, ranging afield from the Nile to the Scottish Highlands and from the Tigris to the Atlas Mountains. War called them to Mesopotamia (193–6), Lyon (197), Parthia and Egypt (197–202), Africa (203), and Britain (208–11). Each time they returned to Rome. Now that they numbered 2000 –

Fig. 4 Septimius Severus' victory speech at Ctesiphon.

like Alexander's horse guard – they made an even stronger strike force.

Of all the heated battles fought by Severus' guard, the one at Lyon in 197 was the deadliest. The British troops of Albinus were as fierce and bloodthirsty as Severus' Pannonians. Severus' own horse was shot from under him – by a lead-weighted dart, says the *Augustan History* – and he sought shelter among the foot. His guard must have suffered dreadful losses of horses and men.

Severus' arch on the Forum in Rome commemorates his Parthian war of 197. On the relief panel commemorating the fall of the Parthian capital Ctesiphon (Fig. 4), the guardsmen, oddly, stand nowhere near the emperor, as he gives his victory speech. They face away to the left and do not seem to be listening. This may be more than an artistic trick to enliven the scene. It may stress that the guard was there as part of the emperor's team not as his audience, looking after his safety rather than minding his words. A coin that shows Galba haranguing troops also has the guard (with the emperor's horse) standing to the side, facing away from the emperor, while the other soldiers look at him spellbound.

Among the horsemen beneath the great pillar is an older man (Plate 5) whom the artist treated with much feeling. His strikingly stern and thoughtful looks foretoken the portraits of troopers on third-century gravestones. His age brings to mind that guardsmen were between 17 and 50 years old. Drawn when the panel was much better preserved than it is today, figure 4 shows Severus also in the group on the right, setting out for home. The emperor's groom (*strator*) brings his stallion from the left, while to the right stands the horse guard, ready to leave. The young soldier furthest to the right, open-mouthed and wide-eyed, strains to bring down his rearing horse. With his hair drilled in curls, his excited bustle, and his youthful looks, he neatly balances the bearded, set features of the older man.

On the way back from Ctesiphon, Septimius Severus, like Trajan, tried to take the desert town of Hatra (El Hadr in northern Iraq) by siege. Hatrene artillery, wielding crossbows that shot two bolts at once, is said to have killed many of the emperor's guardsmen. Severus, like Trajan, may have sent his horse guard to storm the wall. When amidst a rain of burning petroleum they had breached the wall, Severus sounded the retreat, which so angered the guardsmen that the next day they refused to go forward. Again, the town was not taken.[74]

A graveyard at Anazarbos

Not long ago, six gravestones of guardsmen came to light at Anazarbos in Cilicia (near Adana in southern Turkey), from which a great deal can be learned about the horse guard. Since their style and wording are much alike, all six monuments belong to the same period. When emperors stayed in the Orient, their headquarters were at Antioch in Syria, though some units wintered in neighbouring towns so as not to overburden Antioch. Under Hadrian a quarter of the horse guard spent the winter at Gerasa in Arabia; other second-century horse guard detachments stayed at Apamea in Syria, while under the Severan emperors, as the new gravestones show, the horse guard wintered at Anazarbos.

To host the guard was a great honor for a town, burdensome though it might be. In 198 Anazarbos was awarded the highest honor an emperor could bestow: it became the site of a sanctuary for the cult of the emperor, with huge annual games. The town thus was in Severus' good graces and the stationing of the horse guard there was not a hardship but a boon. Good relations between the horse guard and Anazarbos also show in the Greek translation given by two of the six Latin gravestones – clearly, townspeople were meant to read them. One stone gives the name of the unit in Latin as *equites singulares imperatorum*, but in Greek as *Batavi*. Since the Greek version was meant for passers-by, *Batavi* must have been the popular name of the guard. Two further gravestones give the unit's name as *Batavi* even in Latin, a thing never done in the horse guard's own graveyard in Rome.

Very few monuments of the horse guard have been found outside Italy. Soldiers' gravestones were set up only in places at which their units stayed for some length of time. The many gravestones and the good relationship with Anazarbos thus point to a long stay of the guard, more likely during Severus' five years in the Orient between 197 and 202 than during Macrinus' one winter there in 217–18. The choice of Anazarbos as a base makes sense, for as in tactics so in strategy, the guard's place was behind the lines, and Anazarbos lies far to the rear of the border.[75]

Bulla Felix

Hardly had the horse guard come home from the East, when they had to leave with Severus for Africa. There, in 202–3, they may have lent pride and pomp to Severus' appearance in Carthage and in his hometown of Leptis Magna. Perhaps they also saw some fighting in the desert. The only detail known for certain, though, comes from a dedication of 10 June, 203, carved in stone and set up by the supply officers (*curatores*) of the guard in their headquarters room at the New Fort. It gives thanks 'for having come back from this very happy expedition'.

Some quiet years in Italy followed until the brigand Bulla Felix had to be dealt with. The wily robber with his band of six hundred had 'for two years plundered Italy under the very eyes of the emperor and so many soldiers'. He even dared to style his name after the great Sulla Felix. Though hotly pursued, he was never caught, 'thanks to his huge bribes and cleverness'. Corruption, gnawing at the empire's fabric, had reached the guard officers. The emperor 'seethed, for in Britain he was winning wars through others, while he himself in Italy was being worsted by a brigand.' In 207, Severus at last sent 'a tribune of the guard with many horsemen against the robber, threatening the tribune with something awful if he didn't bring in the outlaw alive.' Since he led horsemen, the officer very likely was a tribune of the horse guard. By a stroke of luck he caught the robber in a trap set by a woman. Again the emperor was served well by his horse guard.

In 205 Plautianus, prefect of the praetorian guard, fell from power. Accused of planning to murder Severus and his sons, he was called to the palace where the bodyguard overpowered and killed him. Severus was afraid that if he ordered praetorian guardsmen to seize Plautianus, they might side with the prefect. He therefore posted 'younger members of his personal bodyguard' to lay hold of Plautianus as he entered. Even if Herodian, to whom we owe this detail, made it up, it nevertheless shows how a man who knew the court saw the role of bodyguards: Severus gave them underhanded, grim, and ruthless tasks, some to be done in the palace after midnight. Severus chose younger *equites singulares* since they were less beholden to others. Nero had likewise sent younger, and hence less qualmish, guardsmen to bring in the Pisonian conspirators. Strength also mattered: Nero had his praetorian prefect overpowered by a guardsman 'who was standing nearby because of his great strength'.[76]

During the last years of his life, Severus campaigned in Britain, where in 210 an incident took place that boded ill. Severus and Caracalla rode

> in silence and in parade order to meet the Caledonian army, who wanted to hand over their arms and to discuss terms. When they came to the platform, Caracalla reined in his horse and drew his sword to strike his father in the back. Others who were riding along shouted when they saw this, and Caracalla left off. Severus turned at their shout and saw the sword, but said nothing; rather, he strode out on the stage, accomplished the task at hand and went back to headquarters.

Later he upbraided his son. Caracalla may have answered that he had been rushing to his father's aid, fearing that the groom who helped him off his horse might strike him – that, at least, was to be Caracalla's own fate seven years later. Whatever the truth, the event shows how carefully the emperor's person had to be guarded at all times.[77]

Caracalla and his *Lions*

After Severus' death in 211, Caracalla and Geta, now joint emperors, left Britain. Caracalla sought to kill his brother, but was thwarted for a long time by 'many soldiers and athletes' who guarded Geta on the march as well as in Rome. Some of them, no doubt were horsemen of the guard. Later, when Geta had met his death at the hands of Caracalla's centurions, the emperor took revenge on Geta's guards and followers. He is said to have killed '20,000' men which, if true, would be as many as there were soldiers in Rome. One of those cut down was Iulius Antonius, a training officer (*centurio exercitator*) of the horse guard, whose name was chiseled out on a stone plaque set up

7 A guardsman and his horse on parade: from a gravestone in Rome. The trooper holds shield and spear but wears neither helmet nor cuirass. His hairstyle is that of Septimius Severus.

during Geta's lifetime by the tribunes and drillmasters. Of these four officers, only Antonius seems to have sided with Geta. The horse guard, then, was split in its loyalty to Severus' two sons. Many must have died.

Dio portrays Caracalla as the typical tyrant: like Caligula, Domitian, and Commodus, he killed a hundred bears in the arena and fawned over his soldiers. Unlike Augustus, he did everything he could to prove he was a comrade of his soldiers. He dug, marched, and ate as they did; he caroused with them at drunken parties and went about clad in a silver-embroidered German cape. In the field, he would, like them, forego changes of clothes and baths in order to be closer to 'his men'. They were not likely to murder him, yet he was killed in their midst.[78]

On 8 April, 217 Caracalla set out with the horse guard for the temple city of Carrhae. As he left Edessa (Urfa in south-eastern Turkey) and rode out into the Mesopotamian plain, mother nature called. He bade the party halt and alighted from his horse. When everyone had withdrawn and averted their eyes, he lowered his trousers. Seizing his chance, Iulius Martialis, a praetorian *evocatus* on Caracalla's staff, stepped near the emperor as if to say something important and struck him with a dagger. Quickly returning to his post, Martialis might have hidden his guilt, so the story goes, had he only dropped the bloody dagger. As it was, the weapon gave him away, and one of the Goths of the horse guard ran him through with a spear.

As is often the case, there were others in high places behind the deed. Macrinus, the praetorian prefect, had conspired with Iulius Martialis and with two brothers, Aurelius Nemesianus and Apollinaris, both tribunes of the guard, perhaps the commanders of the horse guard.[79]

Macrinus' motive is clear from the fact that he became emperor in Caracalla's stead. Why the two tribune brothers conspired against the emperor is not known. An inscription shows Aurelius Nemesianus as governor of Mauretania Tingitana during the reign of Severus Alexander (221–35) – a fairly successful career, considering that Nemesianus would have had to hide during the reign of Caracalla's self-styled son Elagabal (218–21).[80]

Iulius Martialis, the *evocatus*, bore Caracalla a personal grudge for

8 Grave-altar of a weapon keeper showing a decorated shield with a barbed javelin and a helmet with a sword.

65

refusing him promotion to centurion. Praetorian *evocati* like Martialis traditionally had very good chances of promotion, but under Caracalla, we are told, other men got the centurionates, men like the Goth who killed Martialis. 'This Goth', Dio says,

> was with the emperor not just as an auxiliary but as a guardsman, for the emperor kept Goths and Germans about him, not only free men but also slaves whom he had taken away from men and women and armed. These he trusted more than the soldiers and honored them, among other things, with centurionates, calling them *Lions*.

The *Lions*, it seems, were regular horse guardsmen, *equites singulares Augusti*, although some may have been chosen not from the *alae* but from irregular units raised beyond the Rhine and the Danube. Under the Severan emperors, units of Frisians, Goths, and other Germans joined the imperial army, and some of their best horsemen may have been promoted to the horse guard. Augustus' class system spurned slave-recruited soldiers, especially in the higher branches of service, hence Dio's reproof. Yet to enlist such men in the guard was not at all new – Labienus did it in Caesar's time. Their promotion to centurions suggests that they were *equites singulares Augusti*, which might account for the envy and hatred of a praetorian *evocatus* who had to compete with them for much sought-after centurionates.[81]

Emperors sought more and more the military backing of ethnic groups: Macrinus had his Mauri, Maximinus his Danubians, Severus Alexander his Orientals, Philip his Arabs. Since regional power bases made the soldiers overbearing, senators and historians frowned upon them, calling them with crass overstatement barbarian or foreign, while in truth they were provincial. This is why Dio and Herodian make Caracalla out to have worn a blond wig and German dress, although many of his 'German' guardsmen were well-romanized citizens. Some came, indeed from abroad, but only in the fourth century did foreigners, such as Galerius' Carpi, dominate the horse guard.

Caracalla, if we may believe Dio, went even further and told ambassadors of the tribes beyond the frontiers to invade Italy if anything happened to him. Rome, he said, would be very easy to capture. This was not idle talk. Frontier tribes might indeed rise when their men or their troops in the imperial service were slighted. The Batavians rebelled in 69, after the *Germani corporis custodes* had been sent home by Galba, and the Mauri rose when under Hadrian their champion Lusius Quietus fell from power.[82]

The horse guard lost no time in striking down Caracalla's murderer. This was as well, for the next emperor, whoever he was, would have had to avenge the murder, if only for his own safety's sake. And he would have had to punish the horse guard if they failed to catch the killer. In 41 the *Germani corporis custodes* had likewise killed the murderers of Caligula, and in 193 Septimius Severus dismissed the praetorians for having failed to avenge the murder of Pertinax. Cassius Dio's remark that Caracalla was murdered in the midst of the soldiers whom he had favored breathes malice, for the horse guard, though unable to prevent the murder, at least avenged it at once. Fulfilling their duty in this way, they served the empire well: their prompt action must have given pause to other would-be murderers.[83]

Macrinus, Elagabal, Severus Alexander

In 218, in a battle near Antioch, Macrinus fought the upstart Elagabal. His horse guard and the praetorians, all of them tall, picked soldiers, fought well as long as Macrinus stayed on the battlefield. But once he fled and they no longer saw the imperial standard on the field, they were ready to give in when Elagabal asked them to be his own guard. This at least is the story told by Herodian. Dio, telling of the same battle, does not mention the horse guard and their strapping size. Perhaps Herodian merely embroidered Dio's account. If he did, it is an indication of how people at the time saw the bodyguard – as tall, fierce, and faithful fighters.

Macrinus' guards fought so well that they turned Elagabal's army to flight. Yet Elagabal restored his lines, Dio says, 'when the men saw him dashing along on horseback, almost godlike, his sword drawn as if to charge the enemy'. Surely he did not do this alone but ringed by an escort, the sight of which may have swayed the outcome the battle. Back in Rome the horse guard rode in Elagabal's great summer-solstice procession honoring the sun god. Herodian singles them out from all other troops of the city-Roman garrison, no doubt since in festive parades horsemen catch the eye.

When in 175 the usurper Avidius Cassius rose against Marcus Aurelius, he may have been slain by his own bodyguard. It was altogether unthinkable that in the first or second century the horse guard would murder an established emperor yet during the third century, when the soldiers were masters of the empire and of the emperors, such murders may have happened. While praetorians killed Pertinax, Geta, Caracalla, and Elagabal, the murderer of Severus Alexander in 235, according to the *Augustan History*, was 'one of the

Germans who held the office of a bodyguard'. The story seems to be made up, though.[84]

The emperor Maximinus, a former horse guard trooper?

C. Iulius Maximinus ruled for three years, from 235 to 238. His extensive warfare on the Rhine and the Danube overtaxed the interior provinces, hence first Africa and then Italy rebelled. Maximinus sent the neighbouring Numidian army to lay Africa waste, while he himself with the imperial field army marched from the Danube across the Alps into Italy. During the ensuing long siege of Aquileia, men of legion II Parthica killed him in his tent. History, written from the senators' point of view, afterwards made the fallen emperor out to be a bloodthirsty, ruthless tyrant, sprung from half-barbarians in 'innermost' Thrace. Inscriptions, however, show Maximinus as a citizen of the respectable town of Nova Italica in Moesia, and modern scholarship rightly sees in him 'a martyr to public duty and a manifestation of the valour of Illyricum'.

Much has been written about Maximinus' career, but facts are hard to come by. Herodian says that because of his size and strength Maximinus joined 'the local army' as a horseman, quickly rose to the rank of provincial governor, and because of his military experience was placed in charge of training all the recruits of the field army in Germany. He put his heart into the task, and the troops, mostly fellow Danubians, becoming fond of him, proclaimed him emperor. To Herodian's vague account the fanciful *Augustan History* adds that Septimius Severus chose Maximinus for his horse guard the day after he joined the army. Is this a playful falsehood to tease the reader or a glimpse of the truth which the author of the *Augustan History* gathered from some source other than Herodian?

A gravestone of the *equites singulares Augusti*, hitherto overlooked, may hold the answer, for it names a 'troop of Iulius Maximinus'. While the names Iulius and Maximinus are very common, their combination is not. No other Iulius Maximinus is known among the horse guard, which makes it likely that this troop leader was indeed the future emperor. The inscription, if it refers to the future emperor, would furnish a framework for the career that led him to the top.[85]

If so, Maximinus joined the army in Lower Moesia as a horseman of an *ala*, the branch of service that took only the tallest recruits and that, in turn, supplied guardsmen to the emperor. Picked for the guard by Septimius Severus, he came to Rome as an *eques singularis Augusti* and there rose to the rank of troop leader (decurion). The career pattern of the horse guard then called for promotion to centurion. Further advances to training officer (*centurio exercitator*) and tribune of the guard, would be in keeping with his earlier rank as troop leader and his later position as governor of Mauretania Tingitana. From beginning to end, Maximinus' career thus was driven by his skill with weapons and training.

Maximinus' striking portrait (Plate 6) shows a stern, strong-willed soldier-emperor. It is the most true-to-life portrait of a (former) horseman of the guard or indeed any Roman soldier. The head, now in the Capitoline Museum in Rome, was broken in antiquity, surely when the senate, betraying the emperor, declared him a public enemy. The no-nonsense look and the short-cropped army haircut pointedly differ from the classical traits and curls of the Antonine and Severi emperors; they stress achievement won by toil rather than birth and education. Since the soldiers held power in the empire, their ideals, too, began to prevail.

As a trooper of lowly origin who rose to high office, Maximinus is but one of many, for under Septimius Severus and Caracalla soldiers took great strides towards becoming the ruling class in the empire. Dazzling careers like Maximinus' depended on how near a man was to the emperor, and service in the guard brought one near to him.

The rise of Maximinus also bespeaks the success of the horse guard as an officer school. Its high standards of recruitment and training turned out inspiring military leaders. Maximinus fought on horseback in the forefront of battle in Germany, killing many of the enemy, and stirring the soldiers to strike bold strokes. He had huge pictures of his deeds painted and set up outside the senate hall so people could see them. A medallion minted in 236–8 shows how the fighting was portrayed: led by Victory and followed by a soldier, Maximinus on horseback rides down two foes. He had learned his skill in the horse guard, and the troopers, surely, were proud that one of their own had become emperor.[86]

The knights of Romulus

In 244, six years after Maximinus' death, Gordian III fell in battle. His successor, Philip, brought the Persian war to a speedy end, but before returning to Rome he took time to visit his birthplace, Shahba, in northern Arabia. He rebuilt the town magnificently and renamed it Philippopolis. An inscription from there mentions a most unusual cavalry regiment of his field army: *ala Celerum Philippiana*.

A thousand years earlier, *Celeres*, the 'Swift Ones', had been the horse guard of Romulus, Rome's founder-king. *Ala Celerum*, therefore, was raised as an emperor's horse guard. On a gravestone found at Virunum in Austria, the widow of an Arab archer calls her husband a troop leader of the *ala Celerum*, 'the finest arrow-shot' (*vir sagittandi peritissimus*). Being highly skilled fighters and, as the inscriptions from Arabia and Austria show, following the emperors on their journeys across the empire, the horsemen of *ala Celerum* not only had the name but also the function of a horse guard.

Turning to the remote past for names and even weapons of army units fits the spirit of the third and fourth centuries. Caracalla, like Alexander the Great, raised a 'Macedonian Phalanx', Diocletian's *triarii* recalled the legions of the Republic, and the *Sabini* and *Latini* of the fourth century harked back to Rome's earliest times. And just as his 'Phalanx' and Macedonian guardsmen proved Caracalla a reborn Alexander, so the *Celeres* proved their founder a reborn Romulus.

Maximinus, rather than Philip, may be the emperor who raised the *Celeres*, for it would have suited his politics to do so. When in January 238 the senate branded him a public enemy, the emperor, according to the *Augustan History*, laid his case before the soldiers, taking the double-dealing senators to task and taunting them with the murders of Caesar and Romulus. Since legend held that the senators had torn Romulus to pieces for having surrounded himself with a *Celeres* horse guard, the raising of such a guard cast the senators in the role of faithless murderers. As a reborn Romulus, the emperor could claim the greater antiquity and hence legitimacy, thus beating the senators at their own game. By guarding himself from their perfidious daggers, he clawed his way to the moral high ground.

A further reason for raising a new horse guard was that, when the senate took the families of the Rome-based horse guard hostage, Maximinus had to fear that his *equites singulares* might turn against him. Wary of his safety, Maximinus may have raised his new horse guard from soldiers not based in Rome. During the second century soldiers in the guard were forbidden to marry, but the slacker discipline of the third century had dropped that ban, and the price for it was now to be paid. Whether or not Maximinus was indeed the founder of the *Celeres* is not certain, for other emperors from 222 to 244, including Philip, cannot be ruled out. What is certain, though, is that Maximinus was murdered by men of legion II Parthica whose families were held hostage by the senate.

Back in the city after the death of Maximinus, the praetorians, and with them no doubt the horse guard, quickly regained the upper hand

70

against the people and the senate. Now that they could safeguard their families they took the offensive. Killing the senatorial emperors Maximus and Balbinus (and sending their newly-hired German guard packing), they found a compromise emperor in Gordian III. The events betray a growing rift between the army and the upper classes that caused much of the crisis of the third century. Maximinus' troops and wars cost huge sums of money which the rich hated to pay, but as the next three decades were to show, the alternative of blundering emperors and untrained armies was costlier still, and grimmer.[87]

After the mid-third century, *ala Celerum*, though part of the emperors' field army, is lost from sight, like most *alae* of the high empire. Its rise, however, between the doubling of the horse guard under Septimius Severus and the unfolding of Gallienus' famous battle cavalry, marks a further step in the growth of the field-army horse during the decades from 193 to 260. Thus, from the time Caesar first raised it, the horse guard kept growing in strength and came to carry ever more weight in battle and politics.

Gallienus' faithful *Batavi*

During the mid-third century, few emperors stayed in Rome for more than a winter. Endless wars, civil and foreign, kept them and their guard forever campaigning. When in 246–8 Philip sought to save Dacia, a field brigade of the horse guard fought by his side. In 250 the *equites singulares* were back in Rome setting up an altar at the Old Fort, but when in 251 Decius fell at Abrittus in Moesia, fighting the Goths, most of his horse guardsmen must have died with him. The survival of the guard itself was threatened by such blows. Though some *remansores* always stayed in the forts in Rome, for the next three and a half decades they left no inscriptions.

Yet the emperors, who had to campaign in person, needed a strong guard. Valerian and his son Gallienus no doubt built the horse guard up to its former strength. In 260 disaster struck again: Shahpur I, emperor of Persia, captured Valerian in Mesopotamia. Shahpur may have taken Valerian by foul play when he came to a parley. In vain Onasander had warned, 'It does not help to think the gods will smite the breaker of oaths – you have to watch out yourself!' Roman field marshals always took their horse guard along to parleys, and Shahpur, who boasted that he caught Valerian 'with his own hands', had the event portrayed on a cameo as a fight on horseback. Valerian's guard thus either fought for him to the death, or followed him into dreaded captivity in Persia. Shahpur used Valerian as a footstool when climbing on to his horse – a high price to pay for keeping one's horse guard either too small or not trained well enough. When Gordian III, Philip, and Decius died in battle between 244 and 251, the best and bravest of the *equites singulares Augusti* no doubt fell with them. Still, Gallienus inherited the guard and its tradition, for *Batavi* are later found among his troops.[88]

When the Goths invaded Macedonia in 268, the threat was such that the tribune of the *Batavi* had to act as governor of the province. During the siege of the capital, Thessalonica, the *Batavi* defended the city against fierce attacks, proving themselves once more the emperor's most effective fighting force. Whatever the horse guard might have done in the past – riding roughshod over civilians, or wresting gold from the emperor – they now made up for, since to save one of the empire's great cities from destruction was a significant achievement. When all was over, the city awarded Valentinus the highest honor it could give, bestowing on him the title 'Founder of Thessalonica'.

Later that year, Aureolus, general of Gallienus' new battle cavalry, rose up as a shortlived usurper. His undoing was that he did not command all of the battle cavalry. Gallienus still had the *Dalmatae* horse and his *Batavi*, now flushed with their victory at Thessalonica. They, and the rest of Gallienus' field army, brought Aureolus down. Again, the guard faithfully fought for the emperor and won.

Gallienus raised new elite corps of horsemen for the field army, the *Dalmatae, Mauri, Sagittarii,* and *Stablesiani.* Among these large corps, the *Singulares* saw their role as a strike force shrink. As the emperor's escort, however, they were closer to the center of power than others and as the most faithful of all troops they were no doubt chosen more carefully and promoted more often.[89]

Galerius' *comites*

Nothing is known of the horse guard for nearly twenty years after 268, a dark time indeed in an ill-starred century. Emperors stayed forever in the field, and disasters abounded for rulers and guardsmen alike. Before the end of the century, however, Diocletian brought back better times, and our sources pick up again. Two altars, set up in 285–6 in the headquarters building of the New Fort in Rome, show that the *equites singulares* still held on to their forts. The guardsmen who took to the field with the emperors thus were drafted from the corps in Rome. Being away from their base for years on end, however, they slowly became units of their own.

In 293–5, a draft of Diocletian's horse guard fought under Caesar Galerius in Egypt. A papyrus calls them *comites dominorum nostrorum*, 'Companions of our Lords' or, to give them their title as it still lives today: 'Counts' (of our Lords). At first *comites* were high-ranking senators who joined the emperor on a campaign as friends and advisors. To win this title, the horsemen must have risen higher than ever before. Their new standing calls to mind the *Celeres* who, fifty years earlier at least in myth had been knights. It also calls to mind a new under-officer rank in the cavalry, *senator*. The horse guards had become aristocrats. Diocletian, it seems, awarded the title *comites* to those who escorted the rulers in the field, distinguishing them from those who stayed in the forts in Rome and bestowing new privileges on them as he did with the horsemen of the frontier *alae* who now joined the field army in privileged units, the *vexillationes*.

A draft of the praetorian horse, under the name *equites promoti domini nostri*, likewise served with Galerius in Egypt. Hence Diocletian and his colleagues, like Caligula, kept a double horse guard of *equites singulares* and praetorians, following the *Doppeltruppen* principle – two units, placed shoulder to shoulder, competing with each other and thus giving their best.[90]

Galerius dwelt in his new capital of Thessalonica most of the time from 299 until his death in 311. There he built a palace and a large, ornate arch with sculptures that seem breathtakingly alive. One scene portrays the emperor amidst his horse guard, heading for the Persian war of 298 giving us a rare picture of the guard in the late third century. The horsemen thronging Galerius' wagon wear the usual third-century dress of long trousers, low-girt, long-sleeved tunic, and a coat fastened by a round fibula over the right shoulder, leaving the sword arm free. No foes seem to be near on this leg of the journey between two cities, for the guards wear neither armor nor helmet. Yet they have not slung their oblong shields to the sides of the horses, but hold them as if danger lurked. If so, they scorned armor when they should have worn it, as do Constantine's men in the battle at the Milvian Bridge (see Plate 20). Their spears, of medium length and with slightly wedge-shaped, barbless blades, seem better suited for thrusting than throwing. In the center of this scene the emperor sits on his open, horse-drawn wagon. He talks to one of the riders, perhaps the tribune of the guard. Between them is a Maurus whose coat and spear are the same as those of the other guardsmen, but whose curly hair pointedly portrays him as a North African. Mauri horsemen had come to Rome with Trajan as colorful outriders (*cursores*), others are known from gravestones of the horse guard, to whose fighting

73

prowess they greatly added as outstanding javelin throwers. Two flags wave near the emperor, perhaps one for the detachment of each unit. The men ahead of Galerius' wagon round out the image of the ruler ringed on all sides by his horse guard, a scene first found on the Aurelian Column, which favored circular compositions.

A stone relief of the goddess Epona, unearthed in Thessalonica (Plate 19), proves that Galerius' horse guard was indeed drawn from the *equites singulares Augusti* in Rome. Carved by the imperial palace workshop, the sculpture must have come from the shrine of Galerius' horse guard, for Epona, one of the most worshipped goddesses of the *equites singulares Augusti* in Rome, had few or no worshippers elsewhere in the East. Her cult, then, was brought to Thessalonica by guardsmen who stood in the tradition of Rome's *equites singulares Augusti*. In 303 praetorians guarded Diocletian in Nicomedia, and so Diocletian and his fellow emperors did not spurn, as has been said, the city-Roman guard and its traditions – indeed, these traditions lived on, unbroken as was the Roman custom, in the new capitals of the empire.

The image of Epona, carved by the imperial palace workshop, bespeaks the guard's new status. In Rome, the guardsmen had to come by their artwork as best they could; now the emperor's artists worked for them. At a time when every soldier could expect ranks and riches, household troops had become part of the imperial aristocracy. The find-spot of the Epona relief shows, moreover, that in Thessalonica the horse guard barracks stood near the palace.

The guard's high status, however, had a dark underside. In his pamphlet on *The Deaths of the Persecutors*, Lactantius, as a Christian, berates Galerius for upholding the religion of his forefathers. After a tirade against the emperor as a sex fiend, Lactantius turns on his horse guard:

> But his companions, too, given such a leader, followed him in rape and freely violated the beds of their hosts. For who would punish them? They took the daughters of the middle class whenever they wished. Noblewomen, whom they could not take, they sought as a favor by the emperor. And one could not refuse the emperor's writ: it was either death or a barbarian son-in-law, for almost all his guardsmen came from the [Carpi] tribe who, at the time of his tenth anniversary, was driven from its homeland by the Goths and had given itself up to Galerius – truly, it was a bane for mankind that those who fled the slavery of barbarians should lord it over Romans.

Even allowing for overstatement and slander, Lactantius illustrates the rule that elite warriors always deem they have rights over others. The Carpi tribesmen who in 303 surrendered to Galerius were Dacian neighbours of the empire. The guardsmen, now as before, came from the border tribes, but the empire had lost the will to romanize them, hence the guardsmen were again as foreign as they had been under Caesar and Augustus. Now, however, the citizens bore the brunt of it: since the guardsmen had become powerful aristocrats, civilians had to fear for their women.[91]

The horse guard of the Later Roman Empire (*scholae*)

In 312 Constantine cashiered the praetorians and the horse guard in Rome and razed their forts to forestall further support for pretenders. Elsewhere, however, horse guard and praetorians lived on in the emperors' field armies, giving rise to new elite classes of units of the line, the *vexillationes palatinae* and *legiones palatinae*. A hundred years later, *comites* still headed the list of cavalry in the field army, keeping their rank above all other units, while another *vexillatio palatina* kept alive the name of the *Batavi*.

The fourth-century horse guard, however, were the *scholae*. It has been said that when Constantine cashiered the praetorians, the *scholae palatinae* took their place. Not so. The *scholae*, being a horse guard, replaced the *equites singulares Augusti*, not the praetorians. Originally there may have been five *scholae*, each 500 strong (the *protectores* perhaps 300):

> *schola scutariorum prima*
> *schola scutariorum secunda*
> *schola armaturarum*
> *schola gentilium*
> *schola protectorum*

The horsemen of the *scholae* thus matched the strength of the 2000 *equites singulares*, not that of the 10,000 praetorian foot-soldiers.

Nor were the *scholae* new. When Septimius Severus allowed the under-officers to gather in clubs (*scholae*), some of these grew so strong that they almost became units of their own. In this way the *scholae* under-officer clubs of the guard became the new units of the guard. A good example is the *schola armaturarum*, a club of drill specialists known among the praetorians of the early third century. The horse guard is likely to have had such a drill team as well. The fourth-century guard unit of the *schola armaturarum* then is the earlier drill team of the horse guard, grown into a unit of its own. Even during the fourth century the *scholae* of the guard were still seen as under-officers' clubs; technically they were not units (*numeri*), and they came under the *magister officiorum*, the secretary for promotions.

When Constantine cashiered the mother units of the praetorians and *equites singulares* in Rome, the *scholae* in his own horse guard and in those of the other emperors lived on. Tasks that had fallen to the horse guard before, now fell to the *scholae*. *Scholarii* horsemen, chosen from the units of the line, now safeguarded the emperor, trained officers, fostered new fighting techniques, heightened the splendor of the court, and served as a strategic reserve. Many again were Germans from the lower Rhine, though now not Batavians but Franks. In every way Constantine's *scholae* thus stood in the stead of Caesar's *Germani equites* and Trajan's *equites singulares Augusti*.[92]

4
TALL AND HANDSOME HORSEMEN

Huge men, unbelievably bold and skilled in handling weapons.

Caesar, *Gallic War* 1,39 on Germans

The armed frontier against the enemy (broken though he may be)
forever hones their homelands in the habit of unflagging toil and
endurance. Life there is all service and even women are sturdier
than men elsewhere.

Panegyrici Latini 9,3,9 on Danubians

Chosen from the frontier *alae*

Augustus' guardsmen, the *Germani corporis custodes*, were 'picked auxiliary horsemen'. At first chosen from tribal cavalry, they were later no doubt drawn from the regular *alae* – at least that was how the emperors of the second and third centuries chose their horse guard. The army's tallest recruits, men six feet and over, enrolled in the *alae*. Horsemen of the auxiliary cohorts, on the other hand, even though they accounted for a third of all Roman cavalry, were barred from the horse guard by their shorter build and lesser skills.

By choosing troopers from the *alae* (or from provincial *singulares*), emperors drew their guard from the best soldiers and at the same time rewarded those who had been bold in battle. This was all the more fitting since old-style battle decorations (*dona*) were not awarded to auxiliaries. Indeed, the system worked so well that Septimius Severus extended it to the legions: no longer bestowing *dona* on outstanding legionaries, he promoted them instead to the praetorian guard.[93]

Most of the horsemen served three to seven years in the *alae* before joining the guard. Their gravestones, therefore, unlike those of other units, show no one younger than 23. Some five years' prior service in the *alae* was enough for the men to be fully trained, to have proven themselves, and still be in their prime. Third-century praetorians likewise joined the guard after four or five years' service on the frontier.[94]

The horsemen thus came to Rome about 23 years old. In the third and fourth centuries, sons of horsemen could join their fathers' units straightaway. This was so in all branches of the army, but for the guard it was a great privilege since the sons did not need to serve beforehand on the frontier. Very likely they trained in Rome as a youth group (*iuniores*). No document mentions such cadets, but the gravestone of a horseman shows his two sons training, one with bow and arrows, the other with throwing-stones (see Plate 14). They would be fit for service at 17.[95]

How were the men picked? Emperors like Antoninus Pius, who never went abroad, may have asked governors of the northern frontier provinces to choose guardsmen from the *alae* under their command – one soldier boasts that Septimius Severus' famous field marshal Ti. Claudius Candidus put him on a promotion list. If most of the men served 20 years in the horse guard, about 70 men were needed every year to keep a unit of 1000 men at full strength. As part of their duty, governors toured the parade grounds of their provinces to check the war-readiness of their troops. There, watching shows and exercises, they could pick outstanding horsemen for the emperor's guard, or grant the troops the right to choose whom they thought best.

Emperors who took the field chose their own guardsmen. Caligula went to Germany to recruit men for his horse guard, or so he said. Trajan, who made a point of knowing every soldier's brave deeds, might choose men for gallantry in battle. War games in which soldiers strove to outdo each other under the eyes of the emperor also helped to find the best. When at Lambaesis the African army paraded its skills before Hadrian, Viator, training officer for the horse guard, stood near the emperor, ready to comment. Septimius Severus, it seems, picked the future emperor Maximinus for his horse guard during military games. In the endless wars of the third century, emperors chose their guard from horsemen of the *alae* serving in the field army. For team spirit, men were chosen in groups.[96]

Boldness, strength, and skill in battle surely were the main criteria for choosing guardsmen, but beauty (*forma*) also counted. Handsome, bright-eyed youths, by and large, rose faster. A Paphlagonian soldier boasts on his gravestone that his youth and beauty no less than his strength had moved Trajan to award him high rank. In the guard, looks mattered even more than in other units since the men were to grace the imperial court. Nero accused the consul Vestinus of keeping 'handsome servants, all of the same age' as a private elite army. Obviously jealous, the emperor felt that he alone should have such young and good-looking escorts.

78

Guardsmen had to be tall, too. The average Roman stood about 170cm (67in) tall, but horsemen for the *alae*, the regular regiments of the line, had to be over 178cm (70in), taller than most modern European royal guards, as tall, in fact as the *lange Kerls* of Frederick William I of Prussia. Since bodyguards must have been taller than the average horsemen of the *alae*, they will have matched Frederick William's *Kaiserkompagnie* as men of over 185cm (73in).

Germans from the Rhine area at first outnumbered all others since they were tall and since Romans liked blond hair. By the second and third centuries, however, other provincials were joining them. Thus Maximinus from Moesia, the future emperor, was picked for the guard because of his strength and speed, his tall, broad, and fine build, and his bright eyes. Tall horsemen frightened the enemy, and since only the fittest could stand the strenuous training, traveling, hunting, and fighting, emperors, in choosing their guardsmen, had to look for the best and could not give in to favoritism or influence peddling. After all, their own life depended on these men. Zosimus rightly called the guard 'the emperor's own, chosen for valor from all others, the most magnificent of all'.[97]

Social origin

The social origin of soldiers in the Roman army is known only in vague outline. Caesar, in the civil wars, welcomed every man, no matter how lowly or criminal, so long as he fought. Under Augustus this changed, and for the first two centuries of the empire Aristides was right in telling the Romans, 'You selected the fittest from everywhere, barring the riffraff, and made these men part of the ruling class'. 'Barring the riffraff' meant that only the better sort of young men were enlisted, those who had a stake in society. Indeed, some legionaries, praetorians, troopers of the horse guard, and even auxiliaries came from the upper classes. They may have been draftees, for Tacitus grumbles that only 'have-nots and drifters' volunteered. Later, when Marcus Aurelius, at a time of crisis, enlisted whomever he could find, including slaves and roving bands of robbers, rabble were no longer kept out of the army and served side by side with men from better backgrounds.

The lowering of recruitment standards not only pleased the well-to-do, who were no longer drafted, but also the 'have-nots and drifters', who found the ever better-paid army a fine job and the hardships of a soldier's life no worse than those of peasant's. In the third century, therefore, there was no shortage of volunteers, although some of them had fled from the law. The hope that the discipline of army life might straighten out such men, failed, however, and undisciplined soldier-mobs came to plunder town and countryside, even Rome herself.[98]

The social origin of the horse guard, known only in five cases, roughly fits this overall picture. Higher classes enrolled during the first two centuries, lower classes during the third. Under Claudius a Cheruscan prince, under Hadrian a municipal aristocrat, under Marcus a Parthian nobleman, and under Severus again a municipal aristocrat served in the horse guard, while under Caracalla some former slaves may have joined. During the fourth century, foreign noblemen enrolled.

Hadrian himself honored the horseman Zenodotus by writing a Greek poem, it seems, for his gravestone. Zenodotus had returned to a town in the Greek East, most likely in Thrace or Syria, where his epitaph praised him in a strikingly aristocratic way:

> This grave of well-cut stone
> hides the body of the great hero, now gone,
> Zenodotos, but his soul has found in heaven,
> with Orpheus, with Plato, a seat worthy of gods.
> A stout horseman of the emperor was he,
> far-famed, well-spoken, godlike. His words
> made him the image of Socrates among Italians.
> To his children he left his father's fair wealth.
> He died a healthy old man, mourned by
> high-born friends, by his town, and by his countrymen.

After leaving service, frontier horsemen often rose to positions of magistrates in their towns, and so, no doubt, did Zenodotus and other horsemen of the guard. The considerable wealth of his father, however, suggests that Zenodotus belonged by birth to the local aristocracy. Like other local worthies he may have joined the army as a common soldier, yet the weight the poem puts on his education, wealth, high birth, and civic spirit, shows that Zenodotos held fast to aristocratic ideals. With such men in the horse guard, it was fitting that second-century troopers rose to the rank of officers and that their sons became knights.[99]

9 *A gravestone from Rome, portraying a youthful, third-century trooper in a long, fringed cape, his tunic girt by a belt with a ring buckle, his sword sheath ending in a disk chape. He flaunts a ring on the left hand, while his horse is decked out with fancy trappings and a braided bridle.*

D   M
PIOVICIO   OEQVI
ISINC CASTRISPRI
XTVRMAC FSINATIO
PANNOnVSMIIITA
TAR   XIIIVIXII
IN   MENSIBVS
IDIFBVSNIROSVERVNI
MEMORIAISIMENIVS
IVIANVSDECFTCFRA
IROTFVSISICETCIPA
D   FSBME

Chosen for their good looks, some guardsmen put on airs and aped the emperors, wearing their hair in the curly Antonine style (Plates 7 and 11). Sons of provincial aristocrats, surely, felt not at all out of place in the horse guard where they might serve together with foreign princes and noblemen. Back home, they took pride in having been one of 'the emperor's own'.

The aristocratic element in the horse guard may have been slight, yet it proves that the *Germani corporis custodes* and the *equites singulares Augusti* were not, as might be thought, near-barbarians. Tacitus had called them outlanders and Dio put down the third-century guard as 'uncouth clod-hoppers and dreadful boors'. Both overstated their point – Romans loved to scorn the not-so-Roman – as did a second-century Italian praetorian who saw fit to boast of his service in the guard by scoffing at *barbarica legio*.

Tacitus' and Dio's slurs against the guard were not just unkind but wrong, for boorish, uncouth guardsmen would embarrass an emperor. This happened to Marcus when he once said something out of the ordinary and no one of those around him understood – they thought he was speaking Greek! Surely Marcus saw to it that this did not happen too often. Slow wits and cloddishness were as unfit for an emperor's escort as were the looks of 'mongrels'. With their wit and dash and the high-born among them, the guardsmen might easily think of themselves as nobility. Galerius even helped them to marry noblewomen.[100]

Ethnic origin

Rome was fatally weakened when, in the fourth and fifth centuries, her army largely made up of foreigners no longer represented the people. Harnessing the fighting power of foreign soldiers, however, had been well worth the price during the first 250 years of the empire and even later, as long as they could be kept under Roman discipline. Troopers from the frontier tribes were much better horsemen than Romans from the inner provinces, for they trained from boyhood on, and in horsemanship training matters more than anything else. In the first and early second century, therefore, the guardsmen came from

10 Broken gravestone of a trooper wearing third-century dress: a long-sleeved tunic, long pants, open shoes, and a cape fastened on the right shoulder. On his left hangs a huge sword with a disk chape.

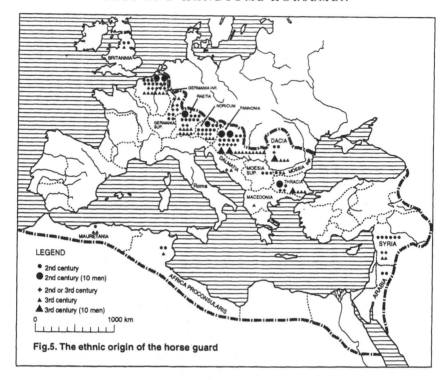

LEGEND

- 2nd century
- 2nd century (10 men)
- 2nd or 3rd century
- 3rd century
- 3rd century (10 men)

0   1000 km

Fig.5. The ethnic origin of the horse guard

the warlike, horse-breeding tribes along the lower Rhine. Of the Tencteri tribe beyond the Rhine Tacitus says:

> Horsemanship . . . is what the boys play, what the young men strive for, and the old ones cling to. Horses are inherited like family, house, and rights, though they are not passed on to the eldest son, like the rest, but to the keenest and best in war.

Customs such as these no doubt lingered also on the Roman side of the river among the Ubii, Baetasii, Batavians, Canninefates, Marsaci, and Frisiavones, from whom the early emperors recruited their horse guard (see Fig. 2).

Beginning with the second century, the emperors drafted guardsmen from all the frontier provinces on the Rhine and the Danube. On the frontiers the warlike spirit lived on longer than in the inner provinces, for Rome kept her frontier people armed. Young men there underwent weapons training, above all the sons of soldiers who often wanted to follow in their fathers' footsteps. Elite soldiers for the guard were the best the outlying provinces had to offer; they were the pride of their homelands and the standard by which Rome judged provincials. On the other hand, few if any troopers hailed from the inner provinces such as Gaul, Dalmatia, Macedonia, or Asia Minor, and none at all from Spain, Italy, or Greece (Fig. 5).

Concerning the horse guard Aristides was therefore right when he told the Romans:

> You thought it unbecoming of your rule that Romans should have to endure military service rather than enjoy the present happiness . . . so you searched the whole empire for those who could do this.

Not all frontier provinces, however, furnished the same number of recruits. A clear pattern emerges from the 210 gravestones of *equites singulares Augusti* giving the homes of the troopers:[101]

Ethnic Origin

Provinces	before 193	after 193
Germany, Britain, Gaul	37 (29%)	10 (12%)
Raetia, Noricum	34 (27%)	11 (13%)
Pannonia	28 (22%)	29 (35%)
Thrace, Dacia, etc.	16 (13%)	29 (35%)
Orient, Africa	11 (9%)	5 (6%)
Total	126	84

A striking change in the guard's ethnic make-up took place between the second and the third century. The share of horsemen from the West (Britain to Noricum) fell from over half to a quarter, while the share from the middle and lower Danube (Pannonia to Thrace) rose from one-third to more than two-thirds. Nothing could better illustrate the sharp shift in the empire's military, and hence political, balance from the Rhine to the Danube. In Domitian's and Trajan's time the bulk of the legions moved to the Danube, but Rome, ever conservative, held on to the tradition of a German horse guard far into the second century, just as it held on to praetorian recruitment in Italy. Then, with the Severan emperors, from 193 on, the make-up of the guard changed drastically. A few Batavians still served, but during the third century seven out of ten troopers came from the middle and lower Danube.

Danubian soldiers, some also German, had outstanding qualities of their own that allowed them to rival the men from the Rhine. As the words quoted at the beginning of this chapter show, it was by their well-known Illyrian spirit of unsparing effort that they matched and outdid German height, horsemanship, and faithfulness. The small number of horsemen from Lower Germany in the third-century guard, however, is due to politics rather than to a waning of the warlike spirit on the banks of the Rhine. Emperors now depended much more on the army, hence the Severi, with their power-base on the Danube, had to draw their horse guard from Pannonia, Moesia, Dacia, and Thrace. From Constantine onwards, the horse guard was again Germanic.

How many horsemen a province furnished to the *alae* had little to do with how many of them joined the guard. In the years from Trajan to Marcus, Lower Germany furnished only 3 per cent of the horsemen in the *alae* but 23 per cent of those in the guard. In the early third century Pannonia likewise furnished only 14 per cent of the horsemen in the *alae* but 35 per cent of those in the guard. Nor did emperors in the second century hire guardsmen mostly from armies they had to humor, for very few guardsmen came from strategically near Upper Germany. During the first 250 years after Caesar, the emperors thus called on riders from the lower Rhine mainly because these were the best horsemen.

Britain, with 6000 men in the *alae*, supplied very few guardsmen. We know of only two in the second century and one in the third. They may be so few because in the second century many of the horsemen stationed in Britain still came from the Rhine, hence their gravestones would not list them as Britons but as Germans. To the third-century Severan dynasty the British army had been an enemy: since they had supported Clodius Albinus in 193–7, British horsemen were not to be honored or trusted.

Some Italians might be expected in the horse guard, for the unit was stationed in Italy and local recruitment was common everywhere – even the praetorians and legion II Parthica had some Italians in their ranks. Yet Italy furnished no horse guard troopers at all since no *ala* was stationed there and the country had long ago lost its tradition of cavalry training.[102]

A few horsemen from the provinces of Africa and Mauretania are always found among the *equites singulares Augusti*. They were chosen for their skills as spearmen, for they could ride faster and throw javelins farther than anyone else. Parthians, Syrians, and Arabs likewise joined the horse guard as specialists. They were outstanding bowmen, and indeed an Arab appears on his gravestone wielding a bow (see Plate 1). Their skills, developed from boyhood on, were such that Hadrian ordered his guard to learn Parthian horse archer maneuvers. Commodus himself trained with Moorish spearmen and Parthian bowmen as his teachers.

Even though Easterners and North Africans furnished one-third of the horsemen in the *alae*, more than nine out of ten guardsmen were Europeans. In its ethnic make-up, the horse guard, then, did not represent the cavalry of the empire. Height played a role. Easterners and Southerners may have shone with valor, skill, and beauty, but they lacked height, being shorter than men from the northern frontier. Nor were the Caesars the only rulers to rely on European horse guards, for Herod of Palestine kept a guard of Gauls and Germans, and, tellingly, also of Thracians, elite troops of such quality that he had to lend them to Aelius Gallus for Augustus' Arabian campaign in 25 BC. They also played a great role in the pomp of the king's funeral. Iuba of Mauretania likewise had Gauls as a horse guard.[103]

The emperors at no time strove to balance the recruitment of the units stationed in Italy. They did not see to it that fleet soldiers came from the East, praetorians from the center, and horse guardsmen from the North – certainly not by the second century, when Thrace supplied horsemen as well as fleet soldiers. Nor did they draw the horse guard from provinces that furnished no praetorians so as to hold the two corps ethnically apart. Noricum, for example, furnished praetorians as well as horsemen to the guard. Augustus, it is true, reserved the praetorium for men of Italian stock, hence during the early centuries horse guard and praetorians were recruited very differently. By the third century, however, the bulk of the soldiers in both corps came from the Danube. Septimius Severus may have wanted his praetorium to embody all the legions, but his hand was forced by the Danubians, his main backers. They claimed the lions' share in both guard corps, and recruitment of the horse guard in Europe grew to 94 per cent.[104]

During the first century most troopers felt that they belonged to an ethnic community such as a tribe or town. Later their loyalty went to a province or a provincial army. Nevertheless, ethnic groups within Roman army units often clung together and set up altars of their own. Among the horse guard, Thracians did this more than others. Yet their altars honor the gods of the horse guard, not those of Thrace. Ethnic fellowship did not turn against the larger community, which is why Hadrian could foster ethnic groupings – of Germans, Dacians, and Raetians – among the horsemen as a means to promote special fighting techniques. Whether or not men of the same tribe, or of the same province, tended to serve in the same troop is not known, although three Frisians indeed served together in one troop as did two Macedonians and two Batavians.[105]

'Peasants from the Illyrian woods, hillmen from the Atlas mountains, and desert riders from the Euphrates valley', ruthless and lost to civilization, ruled the empire from their stronghold in Rome – such was the nightmare of A.v. Domaszewski a hundred years ago. He was right insofar as in the early decades of the third century, when the city of Rome was still the seat of government, the praetorians and the horse guard, in a sense, indeed ruled the empire – they murdered their prefect Ulpian and hounded the consul Dio from Italy. Yet in all this the Danubian ruffians differed little from their forerunners, the second-century praetorians who, though recruited in Italy, stalked the city of Rome with axes. Ethnic character need not be faulted, nor is it safe to say that the breakdown of discipline and the growing ruthlessness since the reign of Commodus stemmed mainly from the emperor's dependence on an army of drifters, frontier peasants, and foreigners rather than of city-dwellers. Perhaps a lack of energetic leadership is also to blame.

At the beginning of the fourth century when foreign troops from beyond the border had become the field army's elite regiments, they entered the horse guard as well. In 303 Galerius had foreigners of the Carpi tribe in his horse guard, and later Constantine named one of his *scholae*-guard units 'The Foreign Regiment' (*schola gentilium*). The horse guard, then, just like the bulk of the horse on the northern frontier, was foreign during the Early Empire, romanized during the second and third centuries, and foreign again during the Late Empire.[106]

Names and citizenship

When a horseman joined the guard he took a Greek or Roman name, or fashioned a Roman name from his native one like the Celtic Ambattus (cognate with English *ombudsman* and German *Amt*). To avoid chaos, enrollment officers saw to it that each trooper took a different personal name. The men, thus shorn of their native tongue-twisters, blended better into the capital. After 193, however, when a Roman veneer (and blending into the city) no longer seemed to matter so much, horsemen bore barbarian names like Durze Mucatra or even doomful Decibal, the name of Rome's worst enemy since Hannibal.

Under Nero, horsemen began adding the emperor's name, Ti. Claudius, to their otherwise single names. This became a tradition. Under Trajan, horsemen who still lacked a full Roman name took the emperor's name M. Ulpius, under Hadrian P. Aelius, and for the rest of the second century T. Aurelius. They did this not to show their closeness to the ruler, for by the early third century the custom came to an end while the guardsmen stayed as close to the emperor as ever. It seems therefore that they took the emperor's name (and passed it on to their freedmen) because he had awarded them citizenship.

Citizens largely served in the legions, non-citizens in the *alae* and cohorts. The latter were often granted citizenship as a reward for valor. A grant of citizenship by Trajan to his horse guard for bravery in the Dacian War may have set the format for the men's Roman names on the rolls of the unit. A newly found diploma of 133, awarding Roman citizenship to an *eques singularis Augusti* upon discharge shows, however, that horsemen did not become Roman citizens just by joining the guard. Perhaps they were given the lesser Latin citizenship, which could also be the reason why so many horse guardsmen bore the emperor's name.

Perhaps one can trace the beginning of this tradition. The horsemen of the guard first earned this honor in 65 for their signal services in the Pisonian conspiracy. Nero relied on the *Germani corporis custodes* in that affair and rewarded his helpers handsomely. A gravestone of the time that names two horsemen Ti. Claudius Chloreus and Ti. Claudius Diadumenus suggests that Nero granted them citizenship. Earlier troopers, such as Gamus, had only single names. Since Nero had awarded citizenship to his *Germani corporis custodes*, Trajan, when he raised his *equites singulares Augusti* thirty years later, may have felt he could grant them no less.[107]

Years of Service

By tradition, Roman soldiers served from 17 to 46 years of age, for the army needed soldiers in their prime, neither too young, nor too old. Enrollment officers, however, might be bribed into enlisting useless bodies who drew pay and benefits. Laws forbade such mischief, and Hadrian 'set the age for the soldiers so that no one would stay against tradition in the camps either younger than manhood demanded or older than compassion allowed'. This standard no doubt applied to guardsmen even more strictly than to others. Yet if we look at the gravestones of the *equites singulares Augusti* in the second and third centuries – those of the first-century *Germani corporis custodes* are too few – we find the following figures:

Recruitment age

age of horsemen	14	15	16	**17**	**18**	**19**	**20**	21	22	23	24	25	26	27
Second-century				1	**9**	**21**	**13**	**20**						
Third-century	1	2	2	**3**	**13**	**9**	**14**	3	3	4	1	–	–	1

Were third-century horsemen recruited at any age? Hardly, for in 14 of the 17 cases falling outside the standard recruitment age of 17 to 20 years, the life-spans of the men are divisible by five. Their age, therefore, was rounded in four out of five cases, which is to say that their true recruitment age remains unknown. On the other hand, among the second-century cases only 29 per cent are divisible by five and among the third-century cases within the standard range of 17 to 20 years, only 21 per cent. Since 20 per cent of all life-spans should be divisible by five anyway, these numbers are far more reliable. In other words, the known recruitment age was between 17 and 20. Most statistics about the recruitment age of Roman soldiers are misleading, for they overlook the factor of age-rounding, but in the case of the horse guard it is safe to say that troopers enrolled, by and large, between 17 and 20 years of age. There is good sense in this, for a recruit who began his training older than 20 could no longer hope to become a top-flight horseman.[108]

Though many have tried, no one has yet ascertained the overall life expectancy in ancient Rome. There are thousands of gravestones, but we are at a loss to know for how many people in each social group such stones were set up. With the horse guard things are better: almost everyone who died during service, even the youngest, was given a gravestone, for pay was high, and the corps had its own cemetery. About 160 gravestones of the *equites singulares Augusti* offer useful data, comparing well with 258 gravestones ten times as many praetorians over twice as long a time. They show that the average age of horsemen who died during service was 34 years. Since they served in Rome roughly from age 23 to 43, and since 34 is near the middle, this means that the mortality rate was even and did not change with the horsemen's age. They died neither particularly young, nor particularly old. If anyone wondered whether, in coming to the great city, many peasant boys from the north died young due to the change in climate or unhealthy living conditions, the answer is clearly that they did not.[109]

And why should they? Chosen for looks and prowess in riding and fighting, they came to Rome as strapping youths in the best of health (see Plate 9). Their forts on the Lateran hill, in the city's finest district, were, like other army camps, built so as to allow fresh air to circulate, they had running water from the nearby Caelian aqueduct and good drainage. Like the frontier forts they may have had their own bath houses (see Fig. 10); if not, the horsemen could use the many baths of the city. For health care they had their own doctors and hospitals. Such conditions were altogether different from those facing Vitellius' army in 69 when it camped, by the tens of thousands, in the bad air on the Vatican field. Many, then, fell sick, and scores, trying to shake off the heat in the Tiber, died from fever. Their deaths prompted others to refuse the offer, however tempting, to join the emperor's guard. Things were altogether different in the horse guard of the second and third centuries. Living conditions were good and injuries from training, rather than diseases, may have been the leading cause of death.

How many guardsmen, then, lived through their full terms of service? Guesses for the praetorians, who served 16 years, run from 54 per cent to 96 per cent. For the horse guard a minimum can be established. From 132 to 145 veterans set up yearly altars, listing on average 24 troopers each. Increased by one-ninth for the three under-officers of each troop, this makes 27 surviving veterans each year. Since a horseman's full term in Rome lasted about 21 years, roughly 70 men had to join yearly to maintain the unit's strength at 1000. Seventy newcomers, set against the 27 discharged troopers, give a survival rate of 39 per cent. That, however, is a minimum. Perhaps some veterans, unwilling to pay for the altars, were not listed; others had been transferred as decurions or centurions to the frontier troops. Hence far more than the minimum of 39 per cent may have lived through their term of service. The survival estimate of 54 per cent for the praetorians seems a fair guess for the horse guard as well (though it is only a guess). If so, while many were crippled in training or fell in war, more than half of those who joined the horse guard at 23 were still around at 43 to see the happy day of their discharge.[110]

When Augustus set the length of service for each branch of the army, he feared rebellion – not because the veterans were kept in service too long, but because after discharge they would still be young enough to go on robbing sprees. He thus set the length of service in the various branches according to social standing and foreseeable behavior of the veterans:

Length of service

praetorians	16 years
legionaries	20 years
horse guard	25 years
auxiliaries	25 years
fleet soldiers	26 years

The only changes afterwards were that legionaries and praetorians also began to serve 25 years once they came from the same social strata as

89

the auxiliaries. The troopers of the horse guard, unlike the praetorians, were not privileged in their length of service, which shows that after discharge they were no less security risks than other auxiliaries. All veterans of the horse guard known from inscriptions served at least 25 years, only those dismissed for a cause, such as age, served less.

Service, even in the guard, was toil (*labor*), and many eagerly longed for their discharge. On their gravestones some reckoned how much time they had done down to days. The recruits of the year 114 proudly note that they were discharged on the very day due, while the cousin of a trooper, bewailing the man's fate in a poem, wrote on his gravestone:

> Just when he hoped to have escaped from this unholy pain
> Pluto sent him, before the deserved day, down to the
> netherworld river.

Here, in the third century, toil, once proudly called *labor* and seen as the ground for Rome's triumphs, has become mere pain, *dolor*. The old Roman discipline is gone.[111]

Careers

A soldier picked for the praetorian guard, was 'transferred' (*translatus*), but if picked for the horse guard, he was 'chosen' (*allectus*) – a term of honor, used for promotion of knights to the senate and for appointment of officers to the emperor's staff. Being thus 'chosen' also meant being raised in rank, for military diplomas call soldiers of the *alae* merely 'troopers' (*gregales*), but they call the horse guardsmen 'horsemen of the emperor' (*equites domini nostri*). Although holders of junior posts in the Roman army were mostly exempt from drudgery work (*immunis*), the horse guard troopers could not have been equally privileged, for if a whole unit was *immunis* who would do the work?

Ambitious horsemen seeking promotion had to win the favor of their tribune, who would then make them part of his staff as *beneficiarii tribuni*. From there they could rise, by imperial appointment, to the ranks of under-officers (*principales*):

> *sesquiplicarius*
> *duplicarius*
> decurion

A decurion led a troop of 30 men. His, surely, was the most coveted position among horsemen, not least since he could count on promotion to legionary centurion, every soldier's dream. Under-officers were better paid, had more room, employed other soldiers as personal followers, and had the right to punish troopers.

Technical ranks in a troop were those of *curator*, who looked after the horses and fodder, *armorum custos*, who took care of the weapons, and *signifer*, who in training and battle carried the standard that every trooper must follow. Though they seem to have formed a group of their own, we do not know what role these three ranks played in the career of a horseman, and whether all three positions had to be held before he could become an under-officer. Nor can we, so far, fit into a career pattern such ranks as flag-bearers, trumpeters, clerical, medical, and training staff.

Hope for promotion upheld the Roman soldier during his long service. Some insight into a man's promotion prospects can be gained from the lists of *curatores* in the New Fort on 1 January, 197 and on 9 June, 203, for in these six years all positions were filled with new men. No one, then, remained *curator* for more than six years, and if a *curator* of April 200 who belongs to neither list is typical, few remained *curator* for more than three years. Promotion in the guard thus seems to have come very fast indeed – and if so, a large number of guardsmen quickly rose through the technical ranks to become under-officers.

An altar of 241 confirms this. It lists 14 men who had all come to the guard from the same *ala*. Five were under-officers (two decurions, three *duplicarii*), three held other ranks, six were privates. Among 14 men only one, at the most two, under-officers might be expected, yet there are five. If this is typical, then nearly half of all horse guardsmen reached the rank of *duplicarius* or decurion. Though for some the promotion came near the end of their service – on their gravestones under-officers average 19 instead of the usual 15 years of service – it was still welcome, both for status and for the cash award upon discharge.[112]

There was even a way for troopers to become officers of the horse guard itself, as did Viator, who saved Hadrian's life. A man could rise from decurion to *centurio exercitator* and serve as training officer in the guard. Or, having become legionary centurion, he could, after a posting with the chief centurions, come back to Rome and make his way through the Rome tribunates. No single career has been found that covered all these steps, but promotions along each step are known from the second century, and under Septimius Severus the future emperor Maximinus 'Thrax' seems to have risen in this way. Since guardsmen were chosen from the finest troopers on the frontiers, since some came from municipal upper-class families, and since they were also highly trained and able to meet the emperor personally, it is easy to see why to them, as to the praetorians, the path lay open to become officers, even during the second century, when such careers were still barred to other auxiliaries.[113]

Not all guardsmen, however, enjoyed equal opportunity in promotions, for the Romans, like the British in India, classed ethnic groups according to how warlike they were. And although Romans knew – by looking at themselves – that 'national character' could change, the national origin of their soldiers mattered to them. Army clerks kept score of the soldiers' home provinces. Writers rated provincials for faithfulness, strength, skill, courage, or discipline. As they saw it, 'unwarlike' Egyptians, 'soft' Asians, and 'slothful' Greeks were not up to the men from Italy, Spain, Macedonia, and Noricum with their 'more respectable appearance and simpler habits'. Germans, though tall, strong, and faithful, 'were easily carried away by fury', unlike the fierce yet slow-witted Pannonians and the doughty, stout men of Thrace.[114]

Such views of national character biased promotions. Of 52 second-century horsemen from Lower Germany, Raetia, and Thrace known from their gravestones, none rose to the rank of under-officer (decurion, *duplicarius*, or *sesquiplicarius*). By contrast, among 45 Pannonians and Noricans of the same period, as many as eight were under-officers. Though the data are scanty, and though it was not impossible for Germans and Thracians to become under-officers, Pannonians and Noricans nevertheless together make up two-thirds of the horse guard's known under-officers. In the praetorium and the fleet, too, Thracians rarely rose through the ranks. Disciplined Pannonians and Noricans thus were deliberately promoted over stout but rude Thracians, backwoods Raetians, and bold but wild Germans. The latter, in Seneca's view, lacked foresight and discipline: 'These bodies, these hearts who know no dissipation, no luxuries, no riches – just give them foresight and discipline and, to say no more, we would have to become true Romans again!'

Nor were the high-born always preferred in promotions, for besides a good reputation, brave deeds (*fortia facta*) were required – a principle that seems indeed to have been followed.[115]

Veterans

Honorable discharges were given every year on 4–7 January. To commemorate the end of their service the veterans set up altars, statues, and plaques, 19 of which were unearthed in the headquarters building of the Old Fort (see Plate 17). They list up to 41 names. Astonishingly, during one celebration the emperor Hadrian rattled off by heart such a long list – although it helped that they all bore his own family name.

Some guardsmen stayed on even after their discharge. Not long ago the gravestone of Silvanus, a veteran, came to light at Anazarbos in Cilicia, where Severus' horse guard spent the winter during a campaign. It would be interesting to know why Silvanus went along on the expedition; perhaps he had a rare skill and was hard to replace. In a similar case, a centurion, perhaps drillmaster of the horse guard, calls himself *ret(entus)*, 'kept on' – clearly in the interest of the unit.[116]

Of the first-century *Germani corporis custodes*, no documents tell us where they settled down after their service. Yet in 68, when Galba cashiered their unit, he sent them back to their own country, perhaps as a punishment. Likewise, when in 193 Septimius Severus disbanded the former praetorians, he banished them for a distance of a hundred miles from the city, and a law banned from the city all soldiers discharged in disgrace. Being sent home thus may indeed have been a punishment for Galba's *corporis custodes*, many of whom might have wished to stay in Rome.

Did the guardsmen of the second and third centuries return to their homelands once their service was over or did they stay on in Rome? Eleven gravestones for veterans have come to light in Rome and nearby country towns, but only two or three in the provinces. On the other hand, five of the veterans' bronze diplomas were found in the frontier provinces but only one in Italy (Fig. 6). Perhaps diplomas were useful only to men who planned to return home and marry non-citizen wives. If so, the distribution of diplomas is not representative, and most guardsmen after their service may have stayed in Rome and Italy, although we cannot say how many. Third-century praetorians and men of legion II Parthica also came from the Danube to Rome, but again we do not know where they retired.

Friends, relatives, or jobs around the camp, might lead discharged horsemen to stay in Rome, quite aside from the lure of the big city. Some, like many praetorians, chose to settle in Italian country towns where, being few and comparatively rich, veterans enjoyed high standing – veterans of the horse guard, like former praetorians,

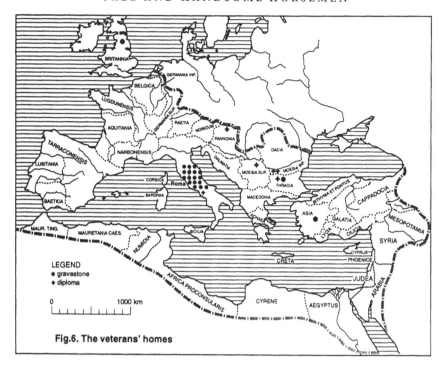

Fig.6. The veterans' homes

proudly called themselves *veterani Augusti*. Those who returned to their homes in the provinces could look forward to civic honors. Lurius, who, as we saw, gained Commodus' help for the tenants on the Burunitanus estate in Africa, was a man to be reckoned with, and so was Zenodotus, father of a family in the Greek East, a man of old money and afterwards mourned by his town. There is nothing strange about veterans returning home after twenty years in Rome. They might still be well known back home, since, for all we know, during their service they may often have come there on furlough, whether to look after their holdings, to see their family and friends, or otherwise to keep alive bonds with their fatherland.

Unlike other time-expired auxiliaries, horse guard veterans did not need to rely on their savings only, for upon discharge they could expect a cash award from the emperor (*commoda*) if the tradition established by Caesar still held good in the second and third centuries. What set them apart most of all, however, may have been a rich store of tales about their service in the capital and with the emperor, to be told and retold during their twilight years.[117]

94

5
ARISTOCRATIC OFFICERS

To Aurelius Valentinus, most excellent
tribune of the Batavi . . . founder of this city.

Marble base from Saloniki, Greece

Duties of a tribune

During the second century, when there was only one fort, a single tribune headed the horse guard. In the third century, after Septimius Severus built the New Fort and raised the strength of the horse guard from 1000 to 2000 men, two tribunes held the command, one over each fort. Their tasks were to look after the discipline, equipment, and training of their troops, and to lead them in the capital as well as in the field.

Aemilius Macer describes a commanding officer's duties as follows:

> The tribunes' task is to keep the troops in the fort, to lead them out for training, to get the keys of the gates, to go round the guards from time to time, to be there when the soldiers get their rations, to check the food, to keep the quartermasters from cheating, to punish crimes within their purview, to be often at headquarters, to hear the quarrels of the men, and to inspect the sick quarters.

Maintaining discipline in the city was of the greatest importance, and for the people of Rome, the tribune's foremost duty may have been guarding the keys to the gates of the fort, as this kept the guardsmen from harassing civilians after curfew. To keep the men out of mischief always was a hard task for Roman officers. They frowned upon free time for the soldiers (*otium castrorum*) and strained to find endless work for their men. Soldiers were all too ready to stray from their camps and become a scourge to civilians, in the city of Rome perhaps even more so than elsewhere. Leaving the fort without permission was therefore strictly forbidden, and climbing over the wall to avoid the watch at the gates called for the death penalty.

In the streets, in parades and shows, the troopers' dress mattered greatly, and Vegetius says of unit commanders, 'Their eagerness and thoroughness are praised when the soldiers go about smartly dressed'. If a unit was not trained well enough and gave ground to the enemy, its commander might be publicly shamed or even cashiered. Yet too much attention to the men provoked the emperor's jealousy; for the same reason the tribune's wife was not welcome on the training ground.[118]

For day-to-day operations the tribune had a large staff. He gave his orders to the chief decurion (*decurio princeps*), who with the help of *librarii* and *adiutores* passed the orders on to the other 30 or so decurions. For supplies and horses there were quartermasters (*curatores*) with their own office in the headquarters buildings; for weapons there were *armorum custodes*, for financial matters standard-bearers (*signiferi*). Training men and horses for battle or for parades was the task of riding masters (*centuriones exercitatores*); archery, it seems, was taught by a *c(ampi)d(octor) s(agittariorum)*. A *medicus castrorum* treated the sick, with an *optio valetudinarii* in charge of the hospital. Tactical orders were transmitted by trumpeters (*tubicines* and *bucinatores*), flag-bearers (*vexillarii*), and standard-bearers (*signiferi*). Orders written on tablets (*tabellae*), perhaps even those given by the emperor, were brought by the *tablifer*, a rank of honor awarded to men in their last years of service.

While the emperors reserved for themselves the right to promote under-officers (*sesquiplicarius, duplicarius, decurion*), tribunes could appoint *beneficiarii tribuni* ('those having received a favor from the tribune'). The men thereby got in line for promotion: no one was discharged in this rank. True to form for an aristocratic society, Roman officers granted promotions as personal favors, not as rewards for achievement, and a number of slots for such *beneficia* was one of every commander's treasured perks. It is a pity we know so little of the duties of the *beneficiarii*. They were at times entrusted with special

11 *Standard bearer, wearing a* paenula-*cloak. The much-decorated standard (the only one known of the horse guard) shows near the top a mural crown, awarded for gallantry during a siege.*

tasks, as when a *beneficiarius* oversaw the dedication of a *schola* room at headquarters. To rise from *beneficiarius* to under-officer, brave deeds were needed. Caesar himself had minced no words in setting true manhood (*virtus*) above mere favor (*beneficium*) as grounds for advancement.

About the tribunes' role in the circus we know nothing, but if Claudius once ordered the praetorian tribunes and even their prefect to join the praetorian horsemen in killing African panthers, the tribunes of the horse guard may also have had to humor the emperor – and the crowds – in the circus. No record of a tribune's role in battle has come down to us. A panel on Constantine's arch celebrating Trajan's victories, however, shows an officer on horseback riding fearlessly by the emperor's side into the Dacian ranks, slaying enemies. Whether he was a praetorian prefect or a tribune of the horse guard, the panel was to proclaim that guard officers fought in the front lines, flinging themselves upon the foe. The tribunes will not have done that very often, for the army needed its officers alive.[119]

Marcus Aurelius and the tribune

Among Rome's finest art treasures are the Aurelian panels, eight magnificent reliefs portraying state ceremonies. Carved for a monument of Marcus Aurelius and later set into the Arch of Constantine, they are well known for their classic style and thorough realism. One of them portrays a tribune of the horse guard (Plate 4), a fact that has hitherto been overlooked and yet casts the meaning of the panel in a new light.

The relief shows Marcus Aurelius in 168, leaving Rome for the Marcomannic war. The emperor (whose head has been replaced) and his chief-of-staff are met by the horse guard at the sacred city limit for the ceremony of setting-out (*profectio*). To the left stand the personified senate and the equestrian order, who have escorted the emperor to this point, while in the foreground lies Lady Flaminian Way who will take the emperor to the north. Waiting to the right stands the horse guard, a group of bearded, helmeted soldiers, one of whom wears a mailshirt and holds a rearing horse.

The officer in the center has so far been mistaken for a standard bearer. Yet he is clearly marked as a knight by the officer's tie around

12 A third-century troop leader (decurion) with three horses, sword, and swagger-stick (virga).

his waist and by his helmet of the classical Greek type. He is therefore the commander of the horse guard, the *tribunus equitum singularium Augusti*. He brings up the white charger, which Marcus is to ride, and holds the imperial battle standard with which the emperor will give the sign for the campaign to begin.[120]

The Aurelian relief not only reveals the tribune's role in the *profectio* ceremony but shows his battle dress as well. He wears knee breeches, a tunic, and a splendid cuirass of scale with scalloped hems. His long, elaborately fringed coat, unlike the emperor's, is clasped not over his shoulder but on his chest. A fairly long sword, clearly a cavalry weapon (not very sharp, the soldiers joked), hangs on a baldric on the officer's right side.

His uniform cost a tribune dear, unless the emperor, as a favor, gave it to him as a gift. At such a price it is small wonder that in civil wars men were keen to kill officers just for their outfits. Indeed, their uniforms so set the knights off from the rank-and-file, that the emperors themselves often dressed as cavalry officers. The pride of an officer was so bound up with his uniform that the prefect of an *ala* whose ill-trained horsemen had given ground to the enemy in battle had his dress slashed and was ordered to stand around at headquarters in the ruined outfit.[121]

The emperor's setting out from Rome (*profectio*) and afterwards his homecoming (*adventus*) marked the opening and the closing of campaigns. These highly symbolic, richly ceremonial events of imperial ideology were broadcast not only on reliefs but also on coins and medallions. One medallion, from 178, shows Marcus and Commodus taking the field, a scene that now may be better understood in light of the panel relief described above: the tribune of the horse guard, wearing an officer's cuirass and wielding a knight's shield (*parma*), awaits the emperors who have already mounted their horses. With the imperial banner, he eagerly leads the way, while a foot soldier trudges behind. To play an outstanding part in state ceremonies was one of the tribune's many duties.[122]

Second-century tribunes

During the first two centuries of the empire, officers had to be well-bred aristocrats, not men risen from the ranks. Knights commanded the auxiliary *alae* and cohorts, while *primipilares*-tribunes commanded the cohorts of the guard, and senators the legions. The horse guard, however, being neither *ala* nor cohort, belonged to the irregular units whose commanders varied widely in rank and title. The emperors, therefore, could appoint to its command whomever they wished.

The rank of the officers regularly commanding the *Germani corporis custodes* is not known, though they may have been *primipilares*-tribunes. Caligula chose lowly gladiators as commanders or training officers of the *Germani*, which is noted by Suetonius mainly because it outraged the Roman sense of dignity. Under Domitian, a knight headed the praetorian horse, the horse guard of the time. In that position he earned the high battle awards of praetorian tribunes and thus, in a way, was their equal in rank. Trajan, then, may have followed precedent when he entrusted his *equites singulares* to *primi-pilares*-tribunes.

Appointing aristocratic tribunes to head the horse guard was a security risk, since knights were more likely than freedmen and gladiators to hatch plots against the lives of emperors. It was good politics, however. In choosing a career officer to lead his guard, an emperor showed the aristocrats that he honored and trusted them. Pliny praised Trajan for this, and for such decisions Trajan became known as 'the best of emperors'.

Tribunates of the Rome garrison were held by *primipilares*, former chief centurions of the legions, the most tried and trusted men in the emperors' service. Unlike knights who commanded the frontier cavalry regiments, tribunes of the city garrison were professional soldiers with many years of service behind them. Moreover, the emperors were familiar with these men from the time they were centurions of the guard, and thus knew whom they could trust and who among them was the right stuff of 'sweat, blood, and deeds' that made the best officers. In giving the command of his horse guard to a *primipilaris*-tribune, Trajan made it clear that his *equites singulares* were as much members of the imperial guard as were the praetorians.[123]

Unlike other tribunes in Rome, tribunes of the horse guard did not serve just one year, but, like commanders of frontier cavalry regiments, held office for three or more years. As cavalry commanders they needed the wisdom that came from years of service. More important, the emperors needed men of the highest skill and trustworthiness in this position and would not lightly part with those who had proven themselves.

Longer service, greater independence, and more power distinguished tribunes of the horse guard from those of the praetorians. They commanded a fort of their own and stood but loosely under the command of the praetorian prefect, while at the praetorian camp nine praetorian and three urban tribunes shared one fortress under the direct command of the prefect. Tribunes of the horse guard met the emperor daily, while praetorian tribunes took turns on the watch at the palace. Nevertheless, horse guard tribunes ranked a step below praetorian tribunes, for the horse guard was an auxiliary unit, while praetorian cohorts were citizen units of venerable tradition.[124]

If, on average, horse guard tribunes served three years, some thirty men held the position before 193. Of these we know no more than six. Their careers, however, are telling, for three of them reached the pinnacle of the equestrian career, the command of the praetorian guard. This proportion is far higher than among praetorian tribunes – the emperors no doubt chose their finest officers as tribunes of the horse guard. And while they could not have earmarked these men so far in advance for a command which they reached only 9, 16, or 21 years later, having commanded the horse guard helped in vying for the post of praetorian prefect. In the fourth century some tribunes of the *scholae* horse guard likewise rose to the rank of second-in-command after the emperor (*magister equitum*). Several tribunes became governors of Mauretania, a position requiring skill in handling large cavalry forces. Even Marcius Turbo, who rose too fast in his career to have time for a governorship, fought as a cavalry commander in Cyrene and in Mauretania. He and a nameless tribune in 120 both earned battle awards. Tribunes of the horse guard clearly had to be battle-proven cavalry commanders.

Commodus, wary of aristocratic officers and trusting no one but his own freedmen, fell back into the bad old ways of Caligula: he raised the former Phrygian slave Cleander to the rank of 'The Dagger' (*a pugione*) that is commander of the watch in the palace or commander of the horse guard. Later Commodus made him praetorian prefect, shoving him along the career path of the horse guard's finest tribunes.[125]

Third-century tribunes

Tribunes in the third century still served about three years, yet they differed from those of the second century in that none rose to the rank of praetorian prefect. Their status was diminished, perhaps, by the fact that there were now two of them, of equal rank, one for each fort. Two of them, Nemesianus and Maximinus, it seems, rose to a governorship in Mauretania.[126]

Third-century tribunes often rose from the ranks, for only there could the emperors find horsemen ready to ride into battle. The sons of senators and knights were not trained well enough to fight on horseback since the skills learned in the *collegia iuvenum* – riding the Troy game – were too ceremonial and stylized to be of much use in the endless wars that followed one another from the time of Marcus onward. Cassius Dio pleaded with the emperors for better training of the sons of the high-born. 'When they come out of childhood into youth (aged 14)', he wrote, 'they should turn their minds to horses and weapons and have paid public teachers in both subjects' – a need that likewise beset aristocratic officers of the Later Middle Ages. Dio shuddered at the thought that men who once served in the ranks carrying firewood and charcoal, could rise to equestrian or even senatorial rank. He had a point, for the rise of an officer class from the ranks, lacking the old aristocratic ethos, worsened the corruption that brought the empire down. Yet the future belonged to those who were willing to fight.[127]

The lowly origin of third-century horse guard tribunes also shows in the way the emperors treated them. Thus in the story, told above, of Bulla Felix, when Septimius Severus sent a tribune of the guard to catch the robber, he 'threatened the tribune with something awful if he didn't bring in the highwayman alive'. Trajan, by contrast, treated his officers with respect – one cannot see him threaten Q. Marcius Turbo with 'something awful' should he fail in a task. With Septimius Severus and his upstart officers, however, such a threat seems fitting. Yet as in the second century, the tribunes' self-esteem knew no bounds. One of them had himself portrayed on a battle sarcophagus dashing headlong, at the emperor's side, into the ranks of the enemy, just like Trajan's guard officer on the Arch of Constantine.[128]

The tribunes of the New Fort seem to have lived in the mansion to the west of the fort, now underneath the apse of St John Lateran (see Fig. 10). The greater part of that noble house had to give way to the fort, but a fine polygonal court with a series of rooms became an appendix to the fort. It would have suited the tribunes very well as a residence, for aristocratic officers maintained an outrageously high style of living even in the field, with servants, baggage trains, baths, and doctors. On the frontiers, commanders of *alae* of 1000 horsemen got 80,000 sesterces a year, 60 times as much as the men, almost like British officers in nineteenth-century India. The tribunes of the horse guard earned no less, but their pay was more in keeping with that of the men, for guardsmen in Rome earned over three times as much as frontier soldiers.[129]

When only part of a unit undertook a task, a centurion, perhaps a *centurio exercitator*, held command with the title of *curam agens* or *praepositus*, a pattern well-known from the frontier auxilia. Some of these *praepositi* counted for almost as much as tribunes. Thus in 246–8, L. Petronius Taurus in a splendid career rose from *praepositus* of the horse guard to the rank of praetorian prefect. He, too, must have been an exceptionally capable officer.

In the mid-third century such *ad hoc* field commands abounded, yet tribunes still served as regular commanders. Under Gallienus when military crisis engulfed the empire, a tribune of the guard at the head of a force that could stave off disaster carried some weight. Aurelius Valentinus, tribune of the *Batavi*, rose to the rank of general (*vir perfectissimus*) and, as acting governor of Macedonia, replacing a man of consular rank, beat the Goths in the siege of Thessalonica.

Though Valentinus is a very common name, the hero of Thessalonica seems to be also the confidant of empress Salonina in a story told by Dio's continuator:

> The wife of Gallienus, hating the hostile looks of Ingenuus, called Valentinus and said: 'I know you to be faithful, and the emperor is right to think highly of you. But not so with Ingenuus – I do not trust him at all. I cannot go against the emperor, but you – keep an eye on the man!' Valentinus answered: 'I hope that Ingenuus will be found true in your service. I, certainly, will not fail in my fealty to your house.'

Ingenuus rebelled in 258. Valentinus, still tribune of the *Batavi* in 268, thus may have headed the horse guard for more than ten years – like Calventius Viator who served Hadrian as a guard officer for 13 years. As tribune of the horse guard and most trusted man in the imperial train, Valentinus, characteristically, kept faith not only with the emperor but, as he said, with the imperial house.

During the later third century all tribunes may have ranked as generals, like Valentinus, for the horse guard grew in importance as the empire came to center more and more on the emperor's person. The tribune who in 286 dedicated a statue of Maximianus at the New Fort certainly held that rank. The last officers to command the two forts in Rome may have boasted the same magnificient title as Constantine's horse guard commander, 'Count of the Godlike Side' (*comes divini lateris*) being forerunners of medieval and modern counts.[130]

6
WEAPONS AND WARFARE

*You mingled the imperial dust and sweat with the troops of
horsemen in training.*

Pliny, *Panegyric* 13 (on Trajan)

Dress

When in 99 Trajan brought the *equites singulares Augusti* to the city, he
saw to it that they looked as Roman as possible. Like frontier troopers
of the line, they wore a sleeveless or short-sleeved linen tunic and
breeches that covered the knees for riding through bushes. They
warded off rain and cold with a hooded, woollen cloak (*paenula*), oval
in shape, with an opening in the middle for the head and two flaps in
front that could be thrown back over the shoulders to free the arms
(see Plate 11). In good weather guardsmen donned a light, piebald,
rectangular cape (*sagulum*), clasped at the right shoulder, but when
riding into battle they wore neither coat nor cape.

For shoes a guardsman wore open sandals (*caligae*) like those of a
legionary: heavy, nailed soles, with straps laced together to above the
ankle. Since the straps almost touched each other, the sandals could
serve as boots; even spurs could be fitted to them. *Caligae* were the
hallmark of a soldier's uniform – as a child in the camp at Cologne,
Gaius used to wear them, and nothing endeared him more to the
soldiers than his nickname *Caligula*, 'bootikins'.

Third-century dress was quite different. A horseman now wore a
long-sleeved tunic and close-fitting, long, tight trousers, held in place,
it seems, by straps under the instep. His shoes were low-cut boots
with open tops; his belt buckle had the shape of a ring, and for a coat
he wore a heavy cape (*sagum*) with embroidered hems (Plate 10).

The new, third-century dress, worn in all branches of the army, guarded a man better against the weather and bushes, thorns, or insects. Caracalla in whose reign the new fashion first appears, wore German dress and shoes when in the field – or so Dio says. The word *caracalla* itself denotes a German cloak. The long-sleeved tunic, long trousers, heavy, embroidered cape, and low-cut boots thus may indeed be German dress, favored by Caracalla as the new army uniform. Caracalla no doubt saw it worn by the guardsmen with whom he took the field. Perhaps he understood that Rome needed not only the northern warriors but also their fighting tactics and hence their equipment. Yet the new dress was not a ruler's whim. It was to stay for centuries.[131]

A horseman's heavy cape (*sagum*) was his pride. In Diocletian's price edict it cost 4000 denarii as against 3000 denarii for a fine war horse. Caracalla's horse guard is said to have liked their capes German style, embroidered with silver. Other dress items, too, were exquisite: a tunic cost twice as much as a finished lion skin and twenty times more than a pair of boots. Red and white belts and red baldrics, also listed in Diocletian's edict, betray smart color schemes. Finger rings, awarded for fighting prowess, were worn on the left hand (see Plate 9).

Dashing dress mattered, and slobs were taken to task by the tribunes. Writers held that the glare of helmets, cuirasses, and swords heightened fighting spirits and struck fear into the hearts of foes. The army therefore insisted on smart looks and shiny weapons, and this may be why the late emperors went so far as to give free weapons and free clothes to all soldiers, saddling the taxpayer with a staggering burden. In a sense, the whole field army became an enlarged guard. The emperor's guard, however, had to stand out from the rest. Galerius gave his horse guard such extravagant uniforms that Lactantius singled them out as the most outrageous items of wanton spending and the cause of the common ruin.[132]

Weapons

A horseman's main weapon was his spear (*hasta*). Gravestones and historical reliefs portray it as not much taller than a man himself – perhaps only because a longer shaft would not fit the frame of the picture. The leaf-shaped blade was meant to pierce shields, cuirasses, or helmets. A javelin (*lancea*), on the other hand, was short and dart-like with a barbed blade (see Plate 8). Since it was lighter and had a strap in the middle of the shaft for the index finger, its range was much longer than that of a spear. In the third century the horsemen, or at least the *protectores* among them, also wielded a lead-weighted spear with a bladeless narrow point (*pilum?*) that could wreak havoc among well-armored horse or foot – such as rebellious Roman soldiers.[133]

The double-edged, pointed, heavy cavalry sword (*spatha*), hung on a baldric. In the first two centuries it was worn on the right-hand side, later on the left. Experiments have shown that swords could easily be drawn from either side. During the second century the sheath ended in a bronze or brass tip, later in a highly decorated, circular chape. With a sword as long as a man's arm, a tall horseman could slash out far to the side. Still, swords were used mainly for hand-to-hand fighting after the spears were gone (see Plate 20), for attacking an enemy's flank, hunting down a fleeing foe, or finishing off the wounded. *Beaux sabreurs*, galloping to the attack with sabers raised overhead, are a modern, rather than a Roman heroic image. Emperors in idealized battle scenes nearly always wield spears. British, German, and Moorish enemies knew few if any swords.[134]

With spears good for thrusting as well as throwing and with swords for close-up fighting the horse guard was neither light nor heavy but medium and hence useful in many different ways. Some guardsmen, mostly Easterners, fought with bow and arrow. On his gravestone such a horseman shows a bow whose tips are curled and angled (see Plate 1). They thus can not have been made of wood and must be bone or antler laths. Very likely his was a composite bow, made of tendons, wood, and horn, glued together. With its short draw and size such a weapon was well-suited for use on horseback.

The horse guard also wielded small battle axes, the double blades of which almost formed a circle and could cut through armor. Some had crossbows built by the guard's own craftsmen (*architecti*), slings hurling rocks of a pound's weight, long lances, or iron *gaesum* spears. A horseman's fighting skills had to be manifold – Egypt's famous fourteenth-century Mamluks likewise trained with sword, shield, spear, bow, maze, and crossbow. Three *tectores* of the horse guard, all in the same troop, are known from an altar dated 250. Perhaps they were not wall-plasterers, as has been said, but heavily-armored horsemen with large shields for defensive tasks such as the parrying of spears.[135]

Shields, about hip-high and made of wood, were shallowly dished and oblong so the rider could protect the horse's head as well. Shield badges may have been uniform, for the badge on Bassus' grave altar (Plate 8) is found on another gravestone, too. One guardsman holds a small, round shield and a bundle of javelins: he specialized in javelin-throwing, perhaps as a *lanciarius* or *gaesatus*.

Mail, and the neckerchief that went with it, are not found on

105

gravestones of the guardsmen, for their reliefs show them unarmored. On Trajan's Column and the Aurelian panels the guardsmen, like most other horsemen, wear mail (see Plate 4). In 312, however, on the Arch of Constantine they are clad in scale (see Plate 20). By that time praetorian guardsmen and emperors alike wore scale, the cuirass of choice because of its lightness. Soldiers hated heavy armor and some donned cuirasses so flimsy or worn that Fronto has a story to tell of how once a commander made a show of ripping the soldiers' feckless armor apart with his bare hands.[136]

A horseman's helmet was a large, one-piece metal bowl with frontguard, cheekpieces, and neckguard (see Plate 8). Arrows and sword-blows could not pierce it, though spears sometimes did. The guard also wore flashy parade armor to overawe the citizens and to lend lustre to the court. Second-century helmets thus could be fitted with crests (see Plate 4), while the gala uniform of the third century (Fig. 7) boasted an eagle-head sword, an eagle-head helmcrest, and leopard or lion skins to deck out the horse. The bird-crested helmets, perhaps gifts from the emperor and worn with gilt armor over red silk, seem to have been used only in parades, not in battle, even though the relief on the Arch of Constantine (see Plate 20) may not be trusted in such matters. For circus shows the horsemen wore highly-wrought sports armor such as gilt copper-alloy helmets with face masks and streamers of blond hair, and bronze chamfrons for their horses.[137]

The saddle, fastened with a girth strap, was of strong wooden construction, padded, and covered with leather (see Plate 14). It was set over a felt blanket and a richly trimmed saddle cloth, while its four horns (*cornicula*) kept the rider firmly seated although he lacked stirrups. Like other equipment, saddles were quality-controlled, but some horsemen padded theirs with down. Under Marcus Aurelius, a stern commander cut such a padded saddle and plucked the feathers from it 'as from geese'.

The leather harness tinkled with bronze pendants, trefoil or crescent shaped. One highly decorative style of harness flaunts a braided bridle and so many leather straps, studded with bronze discs, that they nearly cover the horse's neck (see Plates 9 and 15; Fig. 7). Gallic triplet straps on the side of the saddle, and bronze discs for junction loops, widely used during the first century on the Rhine frontier, are lacking on the reliefs of the *equites singulares Augusti*. Moreover, the breast strap and the crupper are always shown running horizontally as if connected to the girth strap rather than the saddle. The horse guard thus seems to have been outfitted not only with Italian steeds but with Italian saddlery as well.[138]

Fig. 7 Third-century gala uniform.

107

The horse guardsmen, like soldiers of the frontier armies every-where, had to pay for the weapons and their upkeep. The horse guard, therefore, unlike the praetorians, had weapon keepers (*armorum custodes*) in its ranks, whose task it was to rent the troopers weapons for fees taken out of their pay. The well-paid guardsmen no doubt could afford first-class weapons and horses, yet there was a rank order among the branches of the Roman army, and the horse guard's lack of free weapons showed that it ranked beneath the praetorian guard, a distinction that fell by the wayside only in the later third century when all units were given free weapons and the office of weapon keeper vanished.[139]

Horses

During the second century, when it was 1000 strong, the horse guard needed at least 2000 horses if it was to have enough spare animals to replace those lost on marches and in battles. That number rose to 4000 in the third century. Marches from Rome to the theaters of war were long and costly, and losses in battle could be overwhelming, horses being easy targets. Doubling the number of horses thus doubled the guard's staying power.

If the guard had 2000 horses and replaced one-third of them every year, it needed nearly 700 new animals every year. The steeds came from Italy's three famous breeding grounds: Apulia, Campania, and Reate in the Sabine area. Gravestones of *equites singulares Augusti* who died in service have been found there. The men seem to have run stud farms or bought and trained horses for the guard. In Apulia such gravestones date from the second and third centuries, the region therefore supplied horses to the guard from the beginning to the end. Cassius Dio wanted horse races outside Rome to be forbidden so that the best steeds would go to the army.

The horsemen, like the emperor, preferred stallions, as a matter not only of pride but also of fighting prowess, for stallions are keener to bite the enemy and to kick him with fore and hind feet. Although Roman war horses stood on an average only 145cm (57in) high (as against 160cm (63in) for modern ones), warriors took greater pride in bringing down foes with kicks from their horses than with blows from their weapons.[140]

Frontier troopers had to pay fixed prices for their steeds. From 139 to 251, soldiers in auxiliary cohorts paid 125 denarii each, while troopers in the *alae*, who had better horses, paid higher prices, pegged to their pay. The market price for a horse remained one-half of a soldier's yearly pay throughout the centuries, while the soldier's fixed price for a horse did not go up with pay increases and inflation. Over time, this lowered a frontier trooper's outlays for a horse to no more than one-seventh of his yearly pay – a telling yardstick for the rise of the soldier over the taxpayer during the third century. Praetorian horsemen, it seems, got their chargers free. *Equites singulares Augusti*, however, unlike praetorians, had *curatores* in their ranks whom they paid for fodder. If they paid for fodder, they very likely paid for their horses as well.

During the first and second centuries the loss of a horse thus cost a cavalryman dearly, while in the third century it did not hurt as much – unless, of course, he loved his horse. One praetorian, cashiered and banished by Septimius Severus, was so torn by the loss of his horse that when he let the animal go and it kept following and neighing at him, the man, overwhelmed with grief, killed the horse and himself. The troopers of the horse guard likewise loved their animals and took pride in them: of 600 gravestones found at their cemetery in Rome all but a dozen proudly portray horses.[141]

Training

'Cavalry needs boldness and drill', Napoleon said. The emperors saw to both by recruiting fierce and highly trained Batavians and drilling them in all known fighting techniques. High-pitched fighting skills worked doubly well, for would-be foes, seeing Rome's state of readiness, thought twice before going to war. Rome liked to strike her foes with fear rather than weapons, witness Caligula's bridge across the sea and Hadrian's guard swimming the Danube. By their outstanding training alone the praetorians as well as the horse guard, therefore constituted a strategic force.

When war did break out, highly trained troops could be fewer in number and hence be supplied more easily. 'If alone of all other nations we exercise ourselves in peacetime', Josephus says, 'it is so that in wartime we need not contrast our numbers with those of our opponents.' Thorough training, moreover, allowed to strike back swiftly when attacked, as Aristides said: 'You thought that the men chosen from all as the strongest and, above all, as the bravest, ought to train long ahead of time so as to win as soon as they swing into action'.

Knowing that Rome ruled the world because of its thorough weapons training, emperors took a close personal interest in exercises and often went to the parade ground to watch the drills. Domitian, according to Pliny, was a delighted onlooker, and Trajan was so keen an umpire that when he checked weapons for a contest, he tried them for himself. The latter must refer to exercises by the horse guard, for Pliny implies that Domitian should have joined the exercise as Trajan did, and both emperors trained with their guard. The thoroughness of the horse guard's training is borne out by the high number and rank of its training officers.[142]

Training officers

Domitian used 'a little Greek' (*Graeculus magister*) as a training officer, and Pliny scoffs at this – the emperor ought to have chosen a battle-seasoned, highly-decorated Roman veteran. Talking about only one drillmaster, Pliny clearly refers not to the army at large but to the guard. Trajan gave the horse guard Roman centurions for drillmasters. They are known from inscriptions, and Pliny obviously refers to them in his speech. By stressing the Roman-ness of the horse guard's exercises, Pliny may have hoped to forestall those who found fault with Trajan for having brought so many foreign horsemen into the city.

With four *centuriones exercitatores* (some perhaps risen from the ranks of the guard), the horse guard had more and higher ranking drillmasters, and hence higher skills, than any other unit of the Roman army. The four centurions may have trained the guard for dazzling shows in the circus, like the gladiators whom Caligula had put in command of his horse guard. Yet they also had the task of shaping the cavalry tactics of the Roman army. To face a growing threat from heavy, Sarmatian-type cavalry on the Danube and in the East, Trajan and Hadrian tried to raise the fighting skill of the *alae*. New training exercises with new weapons had to be worked out, and the horsemen of the guard, always within reach of the emperor, were best suited for this task. For training officers they needed, and got, centurions who had served on the frontiers and understood the tactical needs of the time.

During the Severan period the horse guard trained under only two *centuriones exercitatores*, one for each fort. Perhaps the guard's training function was dwindling – soldiers, after all, loathed exercises, and the emperors now depended wholly on their good will. In 205, under Severus, Caracalla, and Geta, an inscription named three training officers – surely one for each emperor. With Caracalla bent on murdering father and brother, each emperor needed his own guard with officers he could trust.[143]

110

While bowmen had their own drillmasters, basic, day-to-day training was led by decurions, the troop-leaders. On the frontier, an *ala* of 1000 men, as a rule, had only 24 decurions, each set over 42 men, while 1000 horse guardsmen had 32 decurions, each set over 30 men. The horse guard, therefore, had more officers per man which meant faster promotion, better pay, and better training. Each decurion led his troop (*turma*), named after him, in exercises and in battle. Vegetius says of this under-officer:

> He must be fit, so he can vault into the saddle with his cuirass and with all his weapons, admired by everyone, ride strongly, use his lance keenly, shoot arrows expertly, and teach his men everything needed for fighting on horseback; and he must see to it that they regularly clean and shine their cuirasses, lances, and helmets, for the glare of weapons frightens the enemy greatly, and who would believe a soldier to be warlike whose armor is marred by bad fit and rust? But not only the men, even the horses must be trained painstakingly, hence the decurion has to look after the health and training of the horses as well as the men.

As elite fighters, decurions owned three horses (Plate 12), which allowed them to switch to another horse if the first began to sweat. To enforce discipline they carried a swagger stick (*virga*), a long, straight staff, tapering towards the bottom, with a mushroom-shaped knob atop (see Plate 12). This staff was the auxiliary twin of the centurion's *vitis*-stick, it looked like it and had the same function: to thrash the men. To lift one's hand against it meant harsh punishment, to break it, death.

Decurions often rose to the rank of legionary centurion, some even to the rank of *centurio exercitator* of the guard. As such they still fought in the front line, witness the gravestone of a *centurio exercitator* who fell in battle 'fighting for the fatherland'. The troops admired outstanding training officers so much that in 235 they proclaimed as emperor Iulius Maximinus. A former decurion of the horse guard, and at the time training officer for the field army in Germany, he now became the first 'soldier-emperor'.[144]

Skills to be mastered

Every horseman who joined the guard could jump on to his steed from the left as well as from the right while wearing a cuirass and holding a spear or a drawn sword, but the keenest among them could do this while the horse was running. Even older soldiers and tribunes vaulted into the saddle in one swift move, for to clamber up by throwing one leg over the saddle and then creeping in by dint of heel, knee, and ham, was to lose face. Emperors and army commanders had grooms to lift them up.[145]

Guardsmen trained to fight with every kind of weapon, but above all with the spear (*hasta*). In throwing a spear, a guardsman had to remain seated, unlike the 'little Brits' (*Brittunculi*), as an army intelligence report called the natives of northern Britain. The Elder Pliny, once commander of an *ala* on the German frontier, rhapsodized: 'The cleverness of horses is beyond words. Horsemen hurling spears find that the animals help in difficult throws by swaying their body; also they gather spears lying on the ground and pass them to the rider'. In trying to achieve such rapport, training never ended. A well-flung spear shook and droned as it flew; not so the much shorter, almost dart-like javelin (*lancea*; see Plate 8) with a strap in the middle of the shaft to lengthen its range. Since the index finger held the strap, javelins were hurled more with the fingers than with the fist. Even riding down a foot-soldier required skill, for protected by his shield he would try to stab the horse to make it throw its rider.

Battlefield tactics for large bodies of horsemen needed much practice: riding up to the enemy and firing off several spears before wheeling to the right, rallying, and reaching for more weapons; also lining up for a defensive 'tortoise' formation, or, hardest of all, wheeling to the left and shooting backwards. If assaults were not swift and even, if they failed to keep up a steady, rapid fire on the enemy, they could be broken by attacks against the horses' and their riders' unprotected flanks.[146]

Since detachments of many units fought together in the empire's wars, training maneuvers had to be alike everywhere. The emperors worked them out with the horse guard and published the rules in a handbook, summed up in 136 by Arrian in his treatise on cavalry exercises. After sketching the colorful dress and fancy sports armor of the horsemen, their intricate formations in riding up to the parade ground, their various mock attacks, spear-throwing contests, and lance charges, Arrian continues:

> There follow various shooting exercises with light javelins, or with arrows shot not by their bows but by crossbows, or with

13 Underground strongroom beneath the shrine of the standards in the New Fort. Treasure was stored here, not least the retirement fund of the guardsmen.

stones hurled by hand or by sling against a target set up between the two already mentioned. And here, too, it is a fine thing when the soldiers crush the target with the stones, which is not easy to do. Nor do the exercises end with this, for horsemen ride up with long lances, first holding them straight as for attack, and afterwards as if coming upon a fleeing enemy. . . . These are familiar things to Roman horsemen which they have trained from long ago. But the emperor also ordered them to learn foreign maneuvers, such as the drills of the Parthian or Armenian horse archers, the seesaw attacks-and-flights of the Sarmatian or German lancers, along with manifold ways of shooting that are useful in battle, and the national war-cries: German for German horsemen, Dacian for the Dacians, Raetian for the Raetians . . .

The many fighting techniques marked out here for training could hardly have been meant for any one *ala*, or even for Arrian's own horse guard. Only the Emperor's horse guard had the training and equipment as well as the specialists and ethnic warriors for all this. Tellingly, Arrian reports the last maneuvers not as if he had seen them. The imperial horse guard, recruited from the best fighters in the *alae*, thus set the standards for all the Roman cavalry.

To warrant their own war-cry, the German, Dacian, and Raetian horsemen must have kept up their native style of fighting. The Germans surely were Quadi lancers (*contarii*) raised by Trajan on the Pannonian frontier to fight the Sarmatians and Quadi with their own tactics. Trajan also took Dacian horsemen into the Roman army, not only to drain the manpower of that enemy but, as seen here, to make use of their special fighting techniques. Raetians appear in great numbers in the horse guard under Antoninus Pius. Hadrian must have thought highly of their skill with the iron *gaesum*-spear, for he raised a *cohors Gaesatorum*, 1000 strong, from their militia. Made of heavy iron for 'shooting from afar', a *gaesum* had the same impact as a much larger wooden spear, yet its small size allowed one to take five or more such weapons into battle. Thus, by 136, Germans, Dacians, and Raetians – all from the Danubian frontier – had been brought into the horse guard to hone its 'barbarian' fighting skills.

It further follows from Arrian's treatise that the horse guard gave

14 Family gravestone of Flavius Mocianus. The wife banquets above. One son wields a bow and a target for arrows, the other flings rocks, for both hope to join the guard at age 17.

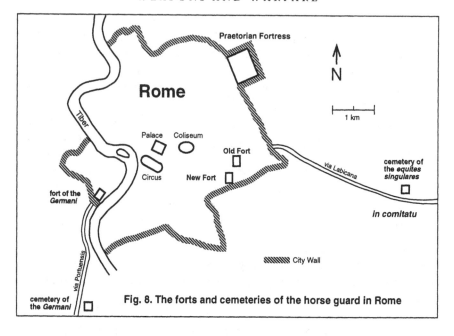

Fig. 8. The forts and cemeteries of the horse guard in Rome

shows in fancy sports armor. Spectacular finds of such armor have been made so far only on the frontiers, not yet in Rome, though emperors like Hadrian loved to give cavalry shows.[147]

Some of the exercises described may have been drilled on the parade ground outside the forts in Rome, the *Campus Martialis Caelimontanus*, now piazza S. Giovanni in Laterano. (If a civilian crossing the *campus* was hurt by a weapon, it was his own fault, the law said.) Proper warlike maneuvers, however, required wider fields, farther away, and to house, train and feed its animals, the guard needed a large estate with stables, barns, meadows, and training enclosures. These fields and grounds, it seems, lay near the unit's graveyard at the third milestone on the *via Labicana*, today's Tor Pignattara (Fig. 8). The area was far enough from the city to be open country and the Maranella river flowed nearby, where the horses could be watered and scrubbed. Exercise was daily, and no doubt one could see a large number of the men riding out early in the morning to the training ground and coming home, tired, around noon. By the early fourth century, a large swath of land there was known to belong to the guard (*in comitatu*). The graveyard of the first-century *Germani corporis custodes* likewise lay far from the city, 2km (1¼ miles) outside the *Porta Portuensis* – perhaps also abutting a training ground.

114

'A general should train the horse by setting up practice battles, pursuits as well as hand-to-hand struggles and skirmishes, both in the plain and around the base of the hills, as far as one can go in broken country.' These words of Onasander are echoed by Vegetius: 'Men who understood military matters laid down that horsemen must unflaggingly train themselves and their horses, not only on level ground but also on steep, rough paths cut by trenches, so they will master everything battle may demand.' For the horse guard broken terrain was not far to seek: it abutted the forts on their eastern side, as can still be appreciated today, looking down from the Lateran terrace to the land below. Finally, for swimming in formation and in full battle gear across wide rivers – one of the horse guard's most outstanding skills – the Tiber offered a fine training ground.[148]

Emperors training with the horse guard

The majesty of the empire, if nothing else, belies the notion that Rome's leading men were dilettantes. Indeed, senators, and among them future emperors, grew up steeped in the skills and wisdom of the men who ran the empire. And they trained in the arts of warfare from boyhood on. Emperors justified their rule in part by their skill and success as field marshals, and many personally took to the field, where they rode on horseback, ringed by their guard. They trained with their horse guard, and if they were to court danger, they must have trained very well indeed.

When an emperor was training, the guardsmen – and at times even the public – watched and judged him. Nero broadcast on coins his skill as a warrior on horseback, and Pliny praised the high standing with the soldiers that Trajan won by his daring deeds on the training field:

> When you mingled the imperial dust and sweat with the horsemen in training exercises, you were like one of them, only stronger and better. As if in real battle you hurled spears from afar and stopped those flung at you with your shield, delighted by the strength of the soldiers and glad when a heavier blow fell on your helmet or your shield. You praised those who struck the blows, you dared them to do so, and indeed they dared!

An emperor thus exercised with his horse guard not only to ensure his safety in the field but also to win the reputation for prowess he needed in his role as *imperator*.

If he trained feebly, an emperor lost face. 'Bad' emperors trained not at all. Domitian, his detractors said, gave up weapons training altogether and rarely rode on horseback in the field. 'To him weapons training was not for the hands, just for the eyes, and not hard work, just fun', they carped, which was to say that he had become no more than an onlooker and no longer trained with his guard as he should have done. This is hardly true, for Domitian went to war often and therefore most likely kept up his training.

A speaker praising Constantine after the battle at the Milvian Bridge in 312, claims that Maxentius, the loser, 'did not venture to the training field, undertook no weapons training, and would endure no dust. In fact', he goes on to say, 'Maxentius was wise not to do so, for thereby those who saw him walk along the marble hallways of the palace would not despise him if he tried to do men's work, he who thought it a field march merely to go to the Sallustian gardens.' In truth, during the battle at the Milvian Bridge, Maxentius and his guard, even though wearing heavy cuirasses, braved the high waters of the Tiber (see Plate 20). They must have practiced this together in field exercises.[149]

Campaigns

Whenever an emperor went to war, his horse guard went with him, a few *remansores* being left behind as caretakers of the forts in Rome. Hyginus, writing when the guard was 1000 strong, reckons with 200, 450, 500, 600, 800 or even 900 guardsmen in the imperial field camp. Very likely, 900 horsemen came along for a campaign, but detachments of various strengths went on special missions with or without the emperor.

In 130, some 240 guardsmen set up an altar at Gerasa in Arabia. On his peacetime travels through the empire Hadrian thus may have taken along only a quarter of his horse guard. On the other hand, in 197 an altar from the New Fort in Rome gives thanks to Hercules for the return of the unit (*numerus*) from the battle of Lyon, hence the whole unit seems to have been there with Septimius Severus. Even older men and veterans went along on campaigns – their experience must have been useful in the field. As a strategic reserve, the horse guard might take part in campaigns with or without the emperor. Thus Tiberius sent the *Germani* with Drusus to quell a revolt of the Pannonian legions.

Twelve gravestones of horsemen who died during campaigns have come to light (Fig. 9). Although they come from the three main frontiers, on the Rhine, the Danube, and the Euphrates, their numbers

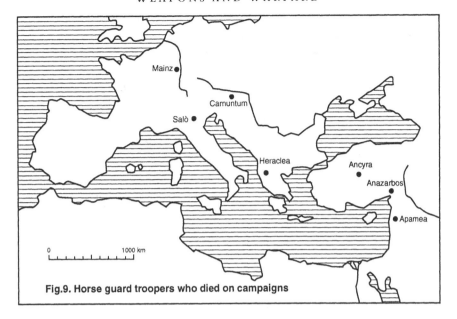

Fig.9. Horse guard troopers who died on campaigns

are few, not because the horse guard traveled little, but because soldiers set up gravestones only where their units stayed for some length of time. In their winter quarters at Anazarbos as many as six gravestones have been unearthed.

The first duty of the horse guard on a campaign was to await the emperor at the city gate for the highly formal ceremony of setting out (*profectio*). There, after the emperor took leave of the senate, the horse guard handed him the imperial banner as a token that the campaign had begun (see Plate 4).[150]

Marches and camps

Since the guard was stationed in Rome, it had to undertake long marches through the provinces to reach the theaters of war. To Aristides, in 144, it seemed that to travel to the frontier, starting from Rome, would take an outward journey of months and years! Cavalry could travel up to 65 km (40 miles) a day without overtaxing the horses. Longer marches, even if less than Caesar's 95 km (60 miles) a day, wore out the hooves of the often unshod horses. Emperors therefore sought water transport for the horse guard. In 48 BC Caesar took his guard to Egypt by ship. Britain or Africa, in the campaigns of

117

Septimius Severus, could be reached only by sea. But even where land transport was feasible, water transport was easier on the horses.

Rowing and steering their ships down the Danube in 101, the emperor and his horsemen cut striking figures on Trajan's Column. River journeys were by no means unusual, however. The frontiers of the empire ran along rivers for long stretches to allow for water transport. Domitian in 89, and then Trajan in 97–8, traveled by ship down the Rhine and the Danube. Domitian, Pliny taunted, 'could not stand the clash and din of oars' and thus had his ship shamefully dragged by another vessel, while, also according to Pliny's vision, Trajan manfully steered and even rowed his own ship. During the Parthian war in 115 'the splashing of the oars and the neighing of the horses each made a louder din than the other', and after sailing down the Euphrates for some length, Trajan had to take the horses overland 'to loosen the stiffness that had befallen them from not being able to stir during the boat trip'.

Among the gods invoked on his altar, one of Hadrian's guardsmen honored the Sea and Neptune, perhaps for help in his never-ending travels, some of them across the sea. For his Parthian war in 197, Septimius Severus took the sea route from Rome to the Orient, casting off from Brundisium. The horse guard must have sailed with him, not only to guard the emperor but to spare the horses and because the sea route was faster, even though the horses might get sea-sick. During the many eastern wars in the third century, warships with billowing sails became familiar sights along the south coast of Asia Minor.[151]

The horsemen, as the emperor's bodyguard, watched over his person throughout an imperial campaign. Nevertheless, in camp praetorian infantrymen pitched their tents next to the emperor's, while the horse guard, together with the praetorian horse, camped further out, lest the noise and smell of the steeds strain the ruler's peace of mind.

Grooms went along on campaigns, one for each trooper, two or more for a decurion, who had three horses. Grooms looked after the baggage train, foraged, and tended the horses. A law of 391 that warns against muddying and dirtying other troops' drinking water gives us a glimpse of the grooms – often lovely boys – and their tasks in that it forbids anyone 'when washing the sweat off the horses, by unseemly nakedness to make the public eye leer'. On gravestones grooms are shown handing spears or helmets to their masters, true forerunners of the pages of medieval knights.[152]

Reconnoitering and parleys

During his campaign on the Rhine, Caligula went reconnoitering with the praetorian horse in a maneuver in which the *Germani* guardsmen played the enemy. In Roman army tradition a field marshal had to see the lay of the land, the enemies, and their weapons, with his own eyes. Reconnaissance by scouts alone was not to be trusted, although there was a regular scouting service (*cura explorandi*) with tried and trusted techniques in which several troops took turns when the enemy was near. The commander had to take some of the risk himself, hence Caligula had to reconnoiter if he wanted to show himself as the strong leader an emperor had to be.

Caligula afterwards bestowed new-fangled scouting crowns on those who had joined him, fair rewards for *exploratores* on whose boldness and skill might hang the fate of a campaign. The crowns awarded for daring daylight reconnaissance were emblazoned with the sun, those awarded for scouting at night with the moon and the stars. The latter was not much safer – Stonewall Jackson in the American Civil War was shot by his own pickets while returning with his staff from a reconnaissance foray on a moonlit night in May 1863. Of scouting crowns nothing is heard later on: perhaps they smacked too much of Caligula. They nevertheless show that great weight was attached to this function of the horse guard.[153]

Reconnoitering before battle was so important that in the first century BC scouts (*speculatores*) became the field marshals' bodyguards. Titus proved his mettle by reconnoitering so recklessly beneath the walls of Jerusalem that he was almost killed and several of his guardsmen lost their lives. Trajan's Column shows the emperor, wearing a cuirass, boldly riding with his horse guards into no man's land where danger lurks. The model for this, as for so many other things, was Alexander the Great who, with his royal horse guard, had daringly reconnoitered as he drew near Gaugamela. On such forays the emperors needed the finest horsemen on the fastest horses, in short, the horse guard. In Caligula's maneuver the praetorian horse escorted the emperor, for the *Germani corporis custodes* played the enemy. On other occasions, surely, the horse guard went reconnoitering with the emperor. The law threatening death to anyone who abandoned his commanding officer (*praepositus*) to the enemy was made, perhaps, for reconnoitering rather than set battles.

To win good terms in a parley, it was essential to impress the enemy. Onasander, writing in the reign of Claudius, found a good-looking horse guard helpful in this:

If the field marshal goes to see the enemy commander face to face, to make or to receive some proposal, he should choose as an escort the strongest and finest-looking of the younger soldiers, stalwart, handsome, and tall men, decked out with magnificent armor, and with these about him he should meet the enemy. For often the whole is judged to be the same as the part one sees, and a field marshal does not decide what to do from reports he has heard, but in fear of what he has seen.

For parleys, then, horsemen were the best escorts. Even if the emperor were seated, a splendid horse guard standing nearby would impress the enemy. Moreover, the safety of the emperor depended on his horse guard's skill and strength. During the parley with Ariovistus, Caesar was guarded by the tenth legion mounted on Gallic horses. Valerian, on the other hand, was captured in 260 by the Persian emperor Shahpur during a parley when his horse guard was overpowered.[154]

Battle

Decisive battles called for the emperor and his guard to be in the midst of the fray, as an inspiration, a shock force, and a reserve. Otho failed his men at Bedriacum in 69 when he and his guard withdrew before the battle: 'With him left a strong force of praetorians, *speculatores*, and horsemen, and those who stayed lost heart'.

An emperor's proper place in battle was in the middle of the second line, as a point of reference to all and out of harm's way. If he planned to lead an attack himself, he stood in the front line towards the right, where the foot in the center linked up with the horse on the wing. His presence was marked by a tall banner (*vexillum*) which he waved for the attacks to begin. Very likely this was the banner presented by the tribune of the horse guard during the setting-out ceremony at the beginning of the campaign and afterwards borne by guardsmen.[155]

Emperors rarely plunged into the fighting, but when they did, the guard had to fight by their side. An idealized battle relief shows Trajan at the head of his horse guard riding down Dacian enemies; Pliny imagined Trajan fighting the Dacian king; and Severus stemmed a rout at Lyon. Third- and fourth-century emperors indeed fought, and sometimes fell, in battle. The main duty of every field marshal, however, was 'to ride by the ranks, show himself to those in danger, praise the brave, threaten the cowards, spur on the slow, encourage the lazy, fill up gaps, replace a company, if necessary, support the wearied, and anticipate a crisis'. In all of this, the horse guard rode with him.[156]

A commander might boost his men's morale if he boldly rode ahead of the troops. Caesar, at Thapsus in Africa, gave the word for the attack, mounted his horse, and cantered headmost towards the enemy line. Like cavalry generals in the nineteenth century, he will have stayed before the troops only for a short while, then reined in as the attack gathered speed and let the spurring horsemen thunder by in their rush toward the enemy. Titus in 68 got the horse guard to fight by telling them he would lead. Elagabal in 218 led his men against the enemy by riding 'almost godlike' ahead with drawn sword. Such gestures were risky, for in order to be seen, the emperor had to ride ahead of the horse guard's protection, and what if his horse bolted and carried him into enemy ranks? [157]

On the battlefield the horse guard, like the praetorians, stood by as a reserve. Holding back reserves was the mainstay of Roman battle tactics; they were needed for surprise attacks, fraught with the risk of heavy losses, or for shoring up the line where it wavered. Marius, even Jugurtha, fought thus in Africa in 112 BC: 'With his horse guard, picked from the sturdiest rather than from friends, he went everywhere, now bringing help to those in trouble and now wading into the enemy where they stood thickest.' Caesar used his *Germani equites* like this at Noviodunum, Titus did the same with his *equites singulares* before Jerusalem, and so did Arrian as governor of Cappadocia with his bodyguard in 135. The pattern is repeated at Lyon in 197, when Septimius Severus' guard saved his right wing at the cost of many lives. Once committed, the guard could be trusted to fight magnificently.

The many different weapons and maneuvers in which the guard was trained allowed the emperor to use them for every kind of surprise and emergency move. Late Roman writers praise self-sufficient, tactically complete units containing a range of specialists, and Hadrian wanted soldiers generally to be trained for every kind of battle. The horse guard had to be the most versatile of all units to fulfill its role as the emperor's last resort on the battlefield.

The horse guard, Tacitus says, was standing by in the rear to swing into action 'as the peak of success, or as support for the floundering'. When their own army began to falter, when men lost heart and tried to flee, it was the guard's dire duty to cut down the cowards. If all was lost, the horse guard covered the emperor's flight.[158]

Swimming rivers

German, and above all, Batavian horsemen, long the core of the horse guard and founders of its traditions, were masters of swimming rivers alongside their horses. They could do this in full armor, while keeping their battle formation. Caesar's horse guard thus swam the Nile, Caligula's the Rhine, and Hadrian's the Danube. Maxentius with his horse guard swam the swollen Tiber. Since there were few bridges in antiquity, this was a highly useful skill. It was all the more valued since Roman soldiers with their heavy weapons dreaded swimming, while German troops, tall, and lightly armed, swam very well.

Even in the Second World War swimming rivers still counted. On June 9, 1940, German cavalry cut Paris off from the north by swimming across the Seine in battle order and under fire. That feat sheds light upon the ancient maneuver (and shows how much the tactics of fifty years ago were still those of Caesar rather than our own). This, in the words of J. Piekalkiewicz, is how they crossed the river:

> When the horses could no longer touch bottom on the shelving bed of the Seine, the riders quickly dismounted, holding the horses' manes with their right hand and the rein with the left. The river was fast-flowing and deep at this point, and the water was cold. Then the current slowed down, the water became shallower, and sandbanks appeared, though the other bank was still far off. The French sentries appeared to have spotted something, and single rifle-shots whipped over the water. From the other bank of the Seine, the squadron responded with a burst of fire to distract the enemy's attention. The horses were swimming well, and the moment they found a foothold, their riders remounted. Three men of the machine-gun squad failed to reach the bank. 'Right, men, up the bank we go!' von Boeselager yelled. It was a critical moment, but on the river bank there was no sign of further enemy movement. The horsemen took the slope in line.

This matches Tacitus' use of *adnare*, 'to swim alongside', and his words 'they steer, at the same time, themselves, their weapons, and the horses'. Cavalry swimming broad rivers was such a spectacular feat that Trajan's Column, when it shows Dacians crossing the Danube, revels in the enemy's losses. In uncontested crossings Roman horsemen never got their arms and armor wet: they loaded them on makeshift rafts which they pulled along as they swam the river with their beasts. The horse guard, however, could cross rivers and inlets under fire since the men kept their weapons ready and, though weighed down by them, stayed afloat by holding on to their horses. Not to be swept away from their steeds in swift flowing rivers, the riders, as the above report shows, swam on the upstream side of their horses.

As they swam, the guardsmen kept their formation so they could face enemies waiting on the other side. It was foolhardy to swim a river where the enemy held a high bank on the other side. Upon reaching a defended river, therefore, the guard scattered, looking for stretches of low bank on the other side. Once they had crossed the deep channel of a river and reached the shallows of the other bank, the first fight would be against enemy archers. When in AD 16 Arminius came with his horse guard to the banks of the Weser, the Romans posted bowmen on their side of the river and withdrew them only when Arminius pulled back his guard. They must have feared that Arminius and his guard would swim across, against which bowmen were the best defense. To meet them, the emperors' guard had outstanding archers of its own, trained to cross rivers in the first line.

Ancient, unchanelled bodies of water offered wide stretches of shallows as battle ground for skilled horsemen. Titus rode with his father's horse guard through the Sea of Galilee to capture Tarichaeae, and the emperor Maximinus, himself a former *eques singularis Augusti*, fought a famous water battle in Germany, ringed, no doubt, by his horse guard.[159]

Crack troops in sieges

In 89 BC, at the siege of Asculum, the consular field commander Pompeius Strabo rewarded a troop of Spanish horsemen so well that there can be little doubt they were among the first to fight their way into the town. Caesar used his horse guard on ships in the amphibious assault on Pharos Island. Likewise, in storming the walls of Iotapata in 67, Vespasian's first attack wave, three men deep, consisted of the bravest of his horsemen, serving on foot, well armored, their spears held at the ready.

Why would a general risk skilled horsemen on a task that belonged to foot soldiers? In taking a fortress bravery mattered more than all else, and horsemen were the daredevils among soldiers. This explains why a standard of the horse guard flaunts a mural crown above three discs (Plate 11), a very high decoration awarded to those who were first over the wall of a besieged city. Guardsmen led such attacks since they were the bravest and since a few hundred bold men often carried a wall. Moreover, as they were the emperor's men, their fame became his, all the more so if the emperor stayed with them during the attack. Tacitus calls it a German custom for guardsmen to attribute their brave deeds to their leader, but it stood the emperors in good stead as well.

Trajan, we are told, sent 'the horsemen' against the walls of Hatra – surely the horse guard. They were hurled back, and the emperor, who stayed with them, was almost killed 'as he rode by'. Elsewhere, too, horsemen rode up to besieged walls – perhaps the speed of their horses allowed them to rush to ill-defended tracts of a wall. To win a mural crown, the horse guard must have taken some other town. Indeed, they were tough street fighters, though under Septimius Severus they failed again before Hatra.

The mural crown proves the guardsmen to have been crack troops in sieges. Tacitus believed that Batavians outdid all others in *virtus*, and there is no reason to doubt their home-bred bravery. Yet Pompeius Strabo's horsemen at Asculum were Spaniards, and Titus' guardsmen before Jerusalem were Easterners. The honor of being picked warriors, ceaselessly trained, and highly paid, surely spurred all horse guardsmen to fearless deeds, whatever their ethnic spirit. Perhaps, then, the Spaniards at Asculum were not just any horsemen but Pompeius Strabo's bodyguard that he kept about himself as a field commander. If so, Rome had a long tradition of putting foreign guards to good use as crack troops in sieges.[160]

Losses

Few reports have survived of the battles the horse guard fought. Jealous frontier troops, however, who felt that service in the guard was not only better paid but also less risky, were mistaken, for the guard often took the field and may well have lost more lives than any other unit. Trajan, to judge from his Column, shared Tacitus' view that there was greater glory in winning victories by shedding non-Roman rather than Roman blood, good grounds for letting the horse guard rather than the legions bear the brunt of fighting. Yet the overriding reason for sending in the horse guard where the slaughter was fiercest was their bravery, witness the many battle-decorations found on the gravestones of praetorian horsemen and legionary horsemen, the latter serving as the horse guard of provincial governors and army leaders.

An Egyptian doctor, in a letter to his parents, gives tantalizing details about such a battle:

> Nobody can go upstream to show his devotion because of the battle of the Anomeritai against the soldiers. Fourteen soldiers of the *singulares* have died besides those of the legionaries, the *evocati*, the wounded, and the missing.

Evocati, soldiers chosen for a second tour of duty, are known as a fighting unit only in the emperor's guard; the *singulares* mentioned here thus seem to belong to the emperor's guard rather than the governor's. If so, the battle took place sometime during the third century when an emperor and his army were in Egypt. The doctor tells of the losses of the horse guard first, and for them alone gives a number, either because he was assigned to the *singulares*, or because their casualties were heavier than those of other units.

Clearly, it was the guard's job to fight, not just to ride in parades or watch over the emperor's life. Their standard bears this out, bristling with battle awards: a gold wreath, three discs, and a mural crown – more than any other auxiliary standard can claim. No individual battle awards (*dona*) to horsemen of the guard are known. Perhaps, like other auxiliaries, they received none. For them, life in Rome was the reward they sought most, and they thanked the gods for their return from a campaign, even if it had been a *felicissima expeditio*.[161]

7
LIFE IN ROME

Rome, where the soldiers wished to live.

Lactantius, De mort. 26f.

The two forts on the Caelius

When Trajan brought the horse guard to Rome in 99, he needed a place to house his 1000 horsemen. Either then, or after his return from the Dacian wars in 102 and 106, he ordered a fort built for them. Two dedications, set up by guardsmen in the fort, prove that they were in their new quarters before 27 October, 113, for on that day Trajan left for the Parthian War, from which he was not to return.

The layout of the fort, later known as the Old Fort (*castra priora*), is unknown. The outline given in Figure 10 is mere guesswork, for only the rooms along the street to the west are known from excavation. The underground strongroom located there and the 'noble hall', with over forty altars, suggest that this was the headquarters building. If so, this was the center of the fort. Rome's praetorian fortress, built in 23, and the New Fort of the horse guard, built under Septimius Severus, both follow standard layout with the headquarters building in the center. The same should be true of the Old Fort, hence its walls must have run further west and north than shown on Figure 10.

Roman army engineers built forts on ground slightly sloping down towards the main gate. The Old Fort lay on the northward slope of the Caelius, hence its main gate very likely was on its north side, looking towards the city as did the gates both of the praetorian fortress and the New Fort. Moreover, the praetorian fortress and the New

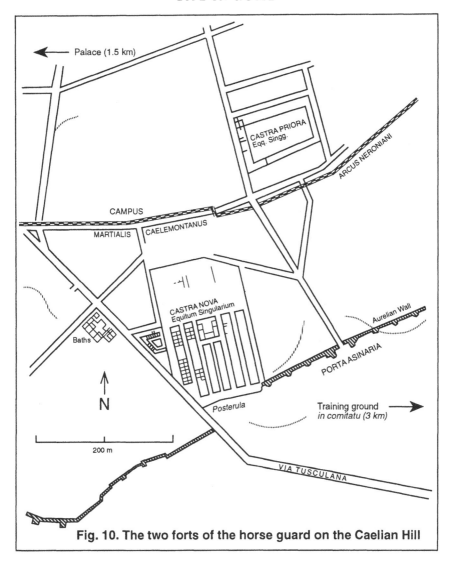

Fig. 10. The two forts of the horse guard on the Caelian Hill

Fort both had defense walls. In 23, Sejanus, as praetorian prefect, had justified building the praetorian fortress by a need to keep the soldiers from getting soft and to control them more strictly with a wall between them and the lures of the city. The Old Fort, then, should have had a defense wall as well, although none has yet been found.

While the fort of the *Germani corporis custodes* stood in the unhealthy area across the Tiber, Trajan placed the fort of his *equites singulares*

Augusti on the cool, breezy Caelian Hill in the wealthy Lateran area. The Neronian Arches supplied the fort with clear, cold water from the Aqua Marcia and the Aqua Julia – a horse may drink 35 liters (9 gallons) a day – while on the city side of the fort there stretched the parade ground and training field of the *Campus Martialis Caelemontanus* (as did the *Campus Viminalis* at the praetorian fortress). Fine marble statues have been found in the fort, a sign of its wealth, and indeed, from here may come the huge bronze statue of Marcus Aurelius on horseback that Michelangelo moved to the Capitolium.[162]

In 193, when he doubled the horse guard, Septimius Severus had another fort built, the New Fort (*castra nova*). A dedication found there shows that the troops were in their new quarters before 1 January, 197. The New Fort is much better known than the old one, for the church of St John Lateran, later rising over it, safeguarded its remains. The New Fort, like the old one, faced north–north–west, towards the city (Fig. 10). The headquarters building in the middle consisted of a main hall, a chapel of the standards, entered by steps of green marble, an underground strongroom for storing treasure (Plate 13), and four offices (*scholae*). The north-eastern office, painted in rose with white lines of geometric decoration, is marked by an inscription as belonging to the *curatores*, the only *schola* in all auxiliary headquarters whose function is known.

In that office, 16 *curatores* dedicated a statue of Minerva. In a fort housing roughly 1000 horsemen, one expects 24 troops, each 42 men strong, or 32 troops, each 30 men strong. The 16 *curatores* thus suggest that in the New Fort there were 32 troops, with 16 stables and 16 barrack blocks each shared by two troops as was the rule in frontier forts. On the frontiers, forts such as this, only 3 ha (7½ acres) in size and with headquarters buildings only 33 × 33 m (110 × 110 ft), housed units of 500 horsemen, not, as here, 1000. In the city, however, forts had to take up less space. The barrack blocks, like those of the praetorians, may have had several stories, and most of the horses were perhaps stabled on the flats south of the hill, outside the later Aurelian Wall. The trapezoidal annex west of the fort seems to have been the tribune's quarter, and the baths across the *via Tusculana* may also have belonged to the fort.

Barrack rooms such as those found to the left of the headquarters building, were 4.6 by 4.2 m (15 by 14 ft) wide, and each housed about

15 Third-century gravestone of a guardsman, showing banquet, bust, and boar hunt, found at Salò near Brescia in Northern Italy. The groom parades his master's eagle-head helmet, while the horse sports flashy trappings and a braided bridle.

seven soldiers. Each of them had, as sometimes in frontier forts, a cellar beneath, and a broad veranda (*propatulum*) outside, on which the soldiers liked to eat. Like other soldiers, the guardsmen, after years of living there, loved their forts as their homes.

The fort wall was 1.3 m (4¼ ft) thick. A large cistern, 4.5 m (15 ft) wide, abutted it near the annex. Part of the wall, towering 7.53 m (25 ft), can still be seen in the apse of the church. A relief on a horseman's gravestone shows how the fort looked from the outside. Its wall, like that of the praetorian fortress, bore battlements with merlons spaced far apart. The gate tower stood out from the wall but did not rise above it, and the two doors closing the arched gateway each ended above in a quarter circle as do fort doors on the Aurelian Column. In 270, the south wall of the New Fort, like the outside walls of the praetorian fortress, became part of the new city wall.

Bronze diplomas reveal that under Severus Alexander and Maximinus both the Old and the New Fort took the name of the ruling emperor: *Castra Priora Severiana* and *Castra Nova Severiana* (or *Maximiniana*). For other units the emperor's name was part of the regimental title; only with the horse guard did it belong to the name of the fort – the two forts somehow, stood for the two units of the corps.[163]

Duties

Guarding the emperor's life was the horse guard's main duty. The *Germani corporis custodes* of the first century were, as their name says, bodyguards, and so, following in their footsteps, were the *equites singulares Augusti* of the second and third centuries. Emperors had a great need for bodyguards. Foreign enemies and civil war rivals sent assassins after them; even in peacetime Augustus had had to thwart many attempts on his life. The oath of the horse guard, therefore, was to hold the emperor's life dearer than all else.

On horseback the guardsmen may have patrolled the garden grounds and outlying areas, but at the palace they no doubt served on foot, hired as they were not only for their riding but just as much for their fighting skills and their faithfulness. On 8 June 69, when his doom gained upon him, Nero fled to the Servilian gardens, escorted by praetorians as well as *Germani*. After a while, the praetorians

16 *Gravestone of a guardsman with spear, horse, and groom. Perhaps he was a Christian, for the inscription fails to call on the Spirits of the Dead. The groom holds the coiled long reins used in warming up the horse.*

withdrew their guard, and later, when he went to his bedroom, Nero found that his bodyguard had left, too. The *Germani corporis custodes*, serving on foot, thus kept watch in the bedroom, while the praetorians stood guard in the outer rooms.

For a guard detail to be ready whenever the emperor wanted to leave the palace, the horse guard must have had a building nearby, such as a hallway (*porticus*), for the men and their horses. At night, too, bodyguards kept watch at the palace, drowsy, no doubt, after a few drinks, and leaning on their spears. Yet this was a most responsible duty and the penalty for leaving one's post was death. Soldiers of the frontier armies, who might see the foe from their tents, nevertheless felt that keeping watch in Rome was an easy task.[164]

Serving as they did by the emperor's side, guardsmen had to groom themselves well, as can be seen on Plates 9 and 11. At home, Batavian warriors wore their hair tied in a knot over the right forehead, something the *equites singulares Augusti* never did; and while no reliefs of the *Germani corporis custodes* have been found yet, surely they likewise had to look smart, just as their names, too, were thoroughly Roman or Greek. Beards were worn by praetorians as well as by horse guardsmen.

To follow orders swiftly and to report correctly the guardsmen had to speak Latin well, though scholars have doubted this. Indeed, first- and second-century inscriptions of the horse guard are as carefully written and as free of mistakes as other city-Roman inscriptions. Emperors also struck notes of personal acquaintance with their bodyguards. Thus Trajan, like every good emperor, knew them by their names and deeds. Hadrian, indeed, wrote epitaphs for them, trying to reach them on the level of poetry, which pleased them, judging from their gravestones. Caracalla, wanting to be their fellow-soldier, went so far as to mix wine for them.[165]

Guardsmen escorted the emperor wherever he went, and crowds cheered – or scoffed – when he passed by. High-handed rulers allowed no one to stand near or even look at them, and the guard saw to it that people kept their eyes fixed on the ground. In the streets of Rome, where even low-ranking officers had a guard of armed men, emperors needed a strong escort. This, it seems, was the task of the *hastiliarii*, known from inscriptions by 135. Scholars thought *hastiliarii* were drillmasters in spear throwing, for the name comes from *hastile*, a wooden spearshaft without blade. Such a weapon, however, being less deadly than a spear, was good not only for training but also for police work. The Aurelian Column (see Plate 3) shows some guardsmen with bladeless staffs and wearing long, fringed coats clasped over the chest like those of the *speculatores*. Though they may have had a flag of their own, we do not know any of their under-officers nor how many they were. Their forerunners, the *speculatores*, numbered 300, Julian in 357 had 200, and Napoleon, with field armies more than twice the size of Roman ones, chose from his horse guard 300 *gendarmerie d'élite*, 'specially selected troopers, responsible for the emperor's safety and the enforcement of his orders throughout the army'. In the third century they were known as *protectores* and wielded a lead-weighted *pilum*-spear, highly effective against heavy-armed men – such as rebellious soldiers.[166]

Not only the emperor himself but other members of the dynasty, too, might be guarded by horse guardsmen. Nero bestowed an escort of *Germani corporis custodes* on his mother Agrippina as an honor, then withdrew it to slight her. To Romans escorts were status symbols, and when Agrippina was stripped of her guard, no one wanted to be seen with her any longer.

As a show of manliness, Roman officers from emperor to cohort commander hunted wild boars. Since the hunt was on horseback, the guardsmen must have gone along, and some of them proudly commemorate boar hunts on their gravestones (see Plate 15). Escort duty also brought horsemen to the emperors' country estates outside Rome. At the imperial villa in the Albani Hills, at Hadrian's Villa near Tivoli, and at seaside resorts along the coast down to Campania, guardsmen stayed often, and long enough, to have gravestones set up there.[167]

The most lavish way Rome knew how to deck herself out was with military pomp, hence a major task of the horse guard was to lend splendor to the capital. At the *profectio*-ceremonies when the emperors left for the field (see Plate 4), at the *adventus* when they returned, also at triumphs and games, guardsmen wore gala uniforms with plumed or eagle-headed helmets and eagle-headed sword pommels, riding horses covered with leopard or lion skins (see Fig. 7). At pay-day parades, every four months, the troopers' uniforms gleamed with gold and silver, while at state funerals they rode splendidly and dashingly around the pyre under the eyes of the emperor, the senate and the knights.[168]

In the circus and the amphitheater the horse guard may have played a very great role, though our sources say little about it. Under Claudius, praetorian horsemen with their tribunes, and even the praetorian prefect, hunted African beasts in the circus. Under Nero, the horse guard speared 400 bears and 300 lions – all in one show. Such feats became common, but we do not know whether the horse guard was in charge of the game park (*vivarium*) or joined the praetorians in the use of the *amphitheatrum castrense* near its forts; nor do we hear anything of its role in the great 'hunts' and cavalry battles the emperors · of the second and third centuries staged to awe the people.[169]

Bowmen among the horse guard greatly strengthened its power to control crowds. Mark Antony rode roughshod over Rome with a band of Ituraean archers, and Cicero bitterly complained about it. Provincial governors, too, used eastern horse archers as guards, for in street fighting they could shoot rioters from afar and at lightning speed. A major task of the Rome garrison was to quell riots. In 180, when Commodus returned to Rome from the Marcomannic War, he addressed the soldiers who had remained in the city and thanked them for their faithfulness. It is not known, however, how many horse guardsmen stayed behind when the emperor took to the field – perhaps not many, since they were the bodyguard.[170]

Praetorians, soldiers of the nearby 2nd Parthian legion, and horse guardsmen got along well with each other, despite some ethnic jealousy. Third-century gravestones often show them as each other's heirs. Whenever civilians and praetorians battled, the horse guard will have fought at the side of the praetorians, with whom they shared not only ties of profession and kinship, but also privileges that made them wish to uphold the status quo. They held each other in high esteem: in the second century a horse guardsman bore the name *Praetorianus*, while in the third century a soldier of the 2nd Parthian legion bore the name *Batao*, 'Batavian Horse Guardsman'.

When the emperor was in Rome, the full strength of the horse guard was not needed for palace duty. Flying columns of horse guardsmen might be sent out to catch robbers besetting Italy, such as Bulla Felix, the scourge of the Italian countryside under Severus.[171]

Life in the city

Since life in Rome was so easy and delightful, every soldier wished to serve there. Indeed, such was the lure of the city that shamefully discharged soldiers were forbidden to come near it. Deserters caught in the city faced the death penalty, while those caught elsewhere were merely returned to service. Yet the city spoiled the soldiers. Men of the Rome garrison were seen, rightly or wrongly, as 'broken by the good life', lazy, easygoing, or tainted by the pleasures of the circus, the baths and the theater, for which ceaseless training and strict discipline were the only cures.[172]

Soldiers liked to pounce on defenseless civilians – overbearing emperors even allowed them to beat down the doors of private houses with axes. And though the law decreed death for disturbing the peace, a military court might well let a marauder from the ranks go free. When horsemen did wrong abroad, governors could write to the emperor, who then might – or might not – punish the men. The wide range of crimes is mirrored by the penalties. Guardsmen might incur beatings, fines, hard labor, demotion in rank, transfer to a lesser branch of service (away from Rome) and, worst, dishonorable discharge, forefeitting all veterans' privileges. No soldiers, only traitors could be condemned to the mines or be tortured. Under-officers could take troopers to task for petty misdeeds; they could also enter a black mark (*nota*) on their record which later might be canceled. The duties of guardsmen called for a stricter discipline than that of other soldiers, hence fewer of them were allowed to marry.[173]

'In their winter quarters soldiers are kept in line by fear and punishment, in the field, promises and cash handouts improve them': this was the wisdom of Roman field marshals. The 'field' for the guard, however, was Rome itself. It was here that the guard fulfilled its duties of guarding the emperor and keeping the provincial armies in awe of his household force. It was in Rome, therefore, that guardsmen had to be paid well; indeed, praetorians received more than three times the pay of legionaries. How much horse guardsmen were paid is unknown. Three times the yearly pay of a horseman in a frontier *ala* would, in the second century, amount to 4200 *sestertii* bronze coins, or 1050 *denarii* silver coins. This was garnished with loot and ever more frequent cash bonuses, one of which, under Caracalla, equalled two years' pay. German guards expected their leaders to give with open hands, hence Caligula heaped so much money upon his *Germani* that their wealth still speaks to us in their tall headstones. The richly-carved gravestones and funerary altars of the second-century *equites singulares Augusti* cost about one-thousand *denarii*, as much as a year's pay, which shows that the guardsmen were wealthy indeed and could afford the luxuries of the city. Frontier soldiers, chafing on miserly pay, could easily believe that the emperors squandered the wealth of Rome on the guard.[174]

To make sure their will would be done, troopers as a rule chose fellow guardsmen for heirs, mostly in the ranks of troop leader, *duplicarius*, standard bearer, or weapon keeper. A powerful under-officer could best see to it that one's last will was fulfilled; moreover, he might treat a trooper better if the man had shown a little 'understanding' and bequeathed him something in his will. The standard bearer was the fiscal officer of every troop; without his goodwill and skill, the bereaved family and friends might not get much of what a horseman had owned. It is a striking fact, and known only from the horse guard, that weapon keepers were named heirs almost as often as all other ranks together. Their duty was to appraise the value of the dead man's weapons and equipment for reimbursement. It paid a trooper to be canny in these matters, for Pliny, who went through the books of several *alae* and cohorts, found there not only carelessness but downright vile greed (*foeda avaritia*). There are no grounds to think that things in the horse guard were any better.

The early *Germani corporis custodes*, however smartly dressed, as armed foreigners in the streets of Rome cannot have failed to frighten the city dwellers. Third-century guardsmen may have looked less foreign than the early *Batavi* but a soldier in the city, even if off duty and unarmed, was easily recognized by his belt, cloak, and short-cropped crew-cut hair (see Plate 10). Lower class third-century praetorians struck city-dwellers as blood-curdling boors, and the growing hatred between soldiers and civilians must have told on the horse guard as well. In the rapport between citizens and guardsmen, as in so many other ways, the second century marks the happiest time of the empire.[175]

Families

Wives may have given soldiers a more fulfilling life and strengthened the army as a social class, but they also lessened the troopers' usefulness. Married men reported for duty less willingly, shunned faraway expeditions, balked when campaigns dragged on, and became desperate when their families were held hostage in civil wars. Of the second-century legionaries in Africa at least 58 per cent were married and perhaps many more. From Hadrian's time on, the African army thus replenished itself with the sons of its soldiers, and the same is true for other provincial armies, but not for the garrison of Rome.[176]

While their number is scant, gravestones set up by wives and children nevertheless give an idea of the marriage rate in the Roman army. Of three major legionary camps on the Rhine, the Danube, and in Africa, data are on hand. They may be compared with the auxilia at large and the horse guard:

Soldiers' gravestones set up by wives and children

Army corps	First century	Second century	Third century
Mogontiacum (Germany)	1%	44%	78%
Carnuntum (Pannonia)	6%	14%	100%
Lambaesis (Africa)		37%	
auxilia generally	5%	17%	29%
horse guard		3%	16%

The above figures give only the minimum marriage rate, since fellow horsemen may also have set up gravestones for married soldiers. Whatever the reason for the low apparent marriage rate of auxiliaries, it is clear that the guardsmen fell even lower. Perhaps second-century guardsmen (like the Popes' Swiss Guard) were not allowed to marry, not even in the half-lawful way that then prevailed. If so, the three known married guardsmen of the time may have had wives before they joined the guard.[177]

By the third century, one out of five horse guardsmen had a wife. Yet compared to the number of married men in the frontier armies, these are few. Perhaps guardsmen were still forbidden to marry. Septimius Severus allowed frontier soldiers to live with their wives and even to buy homes, but it would be in keeping with Roman principles to hold the guard to a stricter discipline, ranking, as it did, even above the legions. If so, the higher proportion of married guardsmen in the third century was due to the fact that more men married when still in the frontier *alae* and brought their wives with them to Rome. Moreover, ten gravestones of guardsmen who died during service were set up by male slaves or freedmen, but only one by a freedwoman. Horsemen thus rarely bought slave girls to make up for their lack of wives.[178]

If in the third century horse guardsmen were still forbidden to marry, they merely matched the praetorians as an elite unit. Why it mattered whether the troops were married or not, became clear at the siege of Aquileia by Maximinus in 238, when soldiers of legion II Parthica killed the emperor and made peace in order to save their wives and children.

Wives, on average ten years younger than their husbands, received gravestones far less often than guardsmen, and mostly cheap ones at that. Yet guardsmen were not strangers to the bliss of married life. One of them is called 'sweetest husband' (*coniux dulcissimus*), another, 'loving', yet another, 'well deserving', while his son is 'sweetest'. In a poem found in the unit's graveyard, a bereft husband, perhaps a horseman, speaks to his wife of his sad life without her:

> . . . we were lucky, then,
> but now you lie sunk in the darkness of the dead,
> and I, among the living, live a sad life.

From the grave, the wife answers:

> Please, forsake the tears and the mourning, dearest husband,
> please, while you live with our offspring.
> Happy to see you again when the Fates call you here
> I will wait for you when you come wandering through the
>   netherworld kingdom.

In funerary banquet scenes on their gravestones (see Plates 2, 9, 10, 12, 15) the quaffing horsemen with their grooms look rather lonely. Sometimes, however, a woman is shown – husbands and wives, obviously, wined and dined together.[179]

Children of guardsmen, like their wives, are known mainly from the third century. Twice sons set up gravestones for horsemen, and twice daughters did likewise. Few though these cases are, they agree with what is known of the *auxilia* at large: daughters set up gravestones as often as did sons, and there are no grounds to think that soldiers cast out unwanted baby daughters. Other family members known from gravestones are mothers, fathers, brothers, uncles, nephews, cousins, even an adopted daughter. Guardsmen and their families thus formed a well-rounded society.

One soldier proudly shows on his gravestone next to himself his two sons training with bow and arrows and rocks (Plate 14) – clearly he hoped his sons would join the horse guard when they came of age. In Roman society, men who strove could rise above their station in life, and across generations upward social mobility was widespread. Parents sought to leave their children well-off, and the son of a horseman could become a knight.[180]

Roman families included the servants. Indeed, during the second century, 'family' to horsemen far more often meant servants than wives and children. A slave boy might cost as little as 200 *denarii*, and a soldier who earned 1050 *denarii* a year could easily pay that much. Grooms exercised the horses, warmed them up on the long rein (see Plate 14), and washed them. Decurions with their several horses had at least two such boys, and a standard bearer also had two, but the army did not want soldiers to keep servants beyond those needed for the horses, since idleness would get the men into mischief. After some years, or upon the death of their master, slaves would be freed, and freedmen could do well, acquiring wives, offspring, and servants of their own. One former groom bought a wife and had a daughter when he was 24, another joined the Night Watch (*vigiles*) to earn a living. While some of the grooms came from Asia Minor or Syria, traditional sources of slaves, others hailed from the same tribes as the horsemen – the Marsacians and Batavians, for example – and had obviously served with their masters in the frontier *alae*.

Some guardsmen loved their boys greatly. One T. Aurelius Sanctinus wrote an epitaph for his 18-year-old boy Miles from the tribe of the Marsacians in Lower Germany. He recommended the boy, 'so obedient and always faithful, the very best, and so dedicated to me, our delight,' to a 'brother', a fellow guardsman, it seems, who had died earlier and was to take care of the boy, 'if the dead are anything'.

Later, when Aurelius Sanctinus was a troop leader (*decurio*) and another of his boys died, this one from the shores of the Black Sea, he gave the two youths a single gravestone with this touching poem:

> Two young men are buried together in this grave
> though born in different places.
> The one buried first
> was the son of the North Sea and the Marsaquian shore.
> The Black Sea and Achilles' land brought forth the other.
> Now, like brothers sprung from the same root
> they are covered by the same weight of Tiburtian stone.

The poem might stand as a motto for the whole graveyard of the horsemen, since all came from faraway lands and were buried there together. Their masters cared so much for the grooms that they are often portrayed them on their gravestones, whether in a conventional scene of bringing up the horse (see Plate 2) or carved for their own sake (Plate 16). Some boys joined their masters in death, dying, it seems, of the same illness and sharing the same grave.[181]

Good-fellowship

Guardsmen who had no families, found company with other soldiers, though not always with their messmates (*contubernales*), whom they chose as heirs less often than might be expected. Army commanders wanted messmates to stay together and looked askance at soldiers who turned elsewhere for companionship. They also mixed young and old soldiers, so the young would learn discipline from the old while the old could use the strength of the young.

The men, of course, spent much time together. As soldiers are wont to brawl, the law laid down the death penalty for wounding a fellow soldier, though if it happened in drunkenness or lust, the penalty could be milder. Sex between soldiers sometimes took place, but those known for it lost face. Some of the city's best known brothels were located on the Caelian Hill, and there were soldiers' pubs as well, for the best loved pastimes, from emperor to simple soldier, were drinking and playing dice. Wine was the bane of the Roman army everywhere, but praetorians and horse guardsmen, well-paid and with far fewer wives, may have wallowed in the camaraderie of the bottle more than others, just as royal guards of the last five hundred years outwenched and outdrank regiments of the line.

Even duty at the palace did not rule out drinking. The killers of Commodus, we are told, sent the emperor's body right through the lines of the drowsy, drunken watch. Caracalla staged drinking bouts with the guardsmen as a way of joining his 'fellow soldiers'. German guardsmen were wont to carouse with their princes, hence Caracalla, who liked them enough to wear a blond wig and German dress, may have taken a cue from them. Serving them drinks he was more likely to win their hearts than was Nero, who regaled hapless guardsmen with his 'heavenly' voice.[182]

8
GODS AND GRAVES

To Jupiter Best and Greatest, Juno, Minerva, Mars, Victory,
Hercules, Fortuna, Mercury, Felicitas, Salus, the Fates, the
Campestres, Silvanus, Apollo, Diana, Epona, the Suleviae
Mothers, the Genius of the Horse Guard, and the rest of the
deathless gods.

Altars of the horse guard

The noble hall of the Old Fort

'I shall never forget the wonderful sight we beheld on entering the
vestibule of the old barracks of the Equites Singulares on the Via
Tasso,' wrote R. Lanciani at the excavation of the Old Fort in 1885.
'The noble hall was found to contain forty-four marble pedestals,
some still standing in their proper places against the wall facing the
entrance, some upset on the marble floor, and each inscribed with the
dedicatory inscription on the front and with the list of subscribers on
the sides.'

Lanciani's 'noble hall' is likely to have been part of the fort's
headquarters, although neither its nor the fort's floor-plan is known.
Had the excavators published an excavation report, we might know
whether the hall was a separate building, a room (*schola*) in the
headquarters building, or perhaps the headquarters' cross-hall (*basi-lica*), the usual place for altars and statues in a Roman fort.

Not all of the hall was unearthed, but the portion that was, yielded
one of the most splendid arrays of Roman army altars ever found. Of
the many, often large, monuments brought to light, the earliest two,
from the time of Trajan, are of travertine limestone, while the others
are of marble. They were all carefully carved and written. Even the
latest altar, set up in 241, shows few signs of third-century decadence.
Most are stately monuments, and none of the shabby relief plaques

139

found in the fort came from here. A particularly well-carved altar shows the god Silvanus looking like an Antonine emperor (Plate 17).

Several inscribed bases from the hall tell of the statues they once bore. Moreover, two full-size statues of gods came to light nearby. One of them, of Pentelic marble, portrays Bacchus and is a first-class work in the style of Praxiteles. Bacchus (*Liber Pater*), not otherwise known to have been worshipped by the guard, was Septimius Severus' home god and may have been honored here for this reason. The other statue represents Mercury. The outstanding beauty of these sculptures bespeaks the wealth – would it be rash to say also the taste? – of the guardsmen.

The 'noble hall' with its niches for statues and its marble floor eventually gave way to a nineteenth-century building, but its inscribed altars are preserved in the National Museum and reveal much about the religion of the Roman army. Knowing that the altars all stood in the same shrine – very likely in the headquarters building – greatly adds to their value. Altars to such unroman gods as the Celtic Toutates Meduris, the Arab Beellefarus, and the Syrian Dolichenus, placed next to the official altars of each year's veterans, show that the Roman army honored all its soldiers' gods. Cicero claimed that Rome owed her greatness to the trust of her people in the gods. He would have felt vindicated had he seen the altars and statues standing at the headquarters of the emperor's horse guard three hundred years later.[183]

The guard's own gods

Nearly all veterans' altars found in the 'noble hall' invoke the same group of 18 deities, as do several altars set up by individual guardsmen to give thanks for promotion or discharge. The gods are the ones quoted in the motto at the head of this chapter. As a group they appear nearly always in the same sequence and after 137 are rounded out by the phrase 'and to the rest of the deathless gods'.

The gods thus gathered became the horse guard's own. The group is met nowhere else save on a series of altars found in Scotland at the fort of Auchendavy on the Antonine Wall. The dedicant, a centurion named M. Cocceius Firmus, clearly wanted to boast of having risen from the emperor's horse guard and did so by invoking the gods of the guard.

Were they Roman or German gods? Jupiter, Juno, and Minerva seem to be the Capitoline triad, duly heading all other gods, while Felicitas, in the middle of the group, is known only as a Roman goddess. If the gods of the horse guard were mainly Roman, the emperor may have chosen them from the state gods worshipped by the Arval Brethren as a means of integrating the horsemen into a thoroughly romanized guard.

On the other hand, the Suleviae Mothers came from Lower Germany, where Augustus and Trajan recruited their horse guard. Soldiers in Rome often clung to their home gods (*dii paterni*), hence the gods of the guard might very well be Celtic or German gods under Roman names. Gatherings of gods were common in Celtic and German religion – witness the dedication by a praetorian from northern Gaul to Jupiter, Hercules, Mercury, Diana (Arduinna), and Mars (Camulus), whom he calls his home gods. Since guardsmen came from many tribes, the list of their home gods grew unusually long. Perhaps then Felicitas was a Celtic or German goddess like Menmanhia, but worshipped by guardsmen under a rare Roman name since better-known names like Fortuna and Salus were already taken. If so, the gods of the guard were gods of Lower German tribes, freely chosen by veterans for their altars. Later, when one such list of gods became the model for other altars, that group of gods became the guard's own.[184]

Celtic-Germanic goddesses

Since Caesar made Gaul Rome's main recruiting ground for horse-men, Epona, the Gallic goddess of the horses, was worshipped all along the north-western frontier. The horse guard too honored her highly. By the second century, a new cult image, the 'imperial type', showed her seated between horses. Though not a mother herself, Epona had much in common with the Mother Goddesses and, like them, could manifest herself as a group of goddesses (*Eponae*).

Monuments and inscriptions for Epona are known from England to the Balkans, but not from the Greek East. Only at Thessalonica has a relief of the goddess come to light (Plate 19) made by the same workshop that carved Galerius' arch in Thessalonica in 299–303. The style and uniqueness of the relief suggest it was made for the emperor's horse guard, perhaps for the barracks at or near the new palace. The rather large relief, whether paid for by the emperor or by the men themselves, contrasts with the much poorer, almost home-made relief plaques from the Old Fort in Rome. Clearly the guard was rising in status as the emperors' household troops became a military aristocracy in the Later Empire. By sponsoring such a fine cult relief,

Galerius, good heathen that he was, upheld Rome's hallowed traditions against the newly spreading, unroman cult of Christ.[185]

Among Celts and Germans, triple mother goddesses watched over people and places. Thus Matronae Domesticae cared for the home, Ambiorenenses for people on both sides of the Rhine, Campestres for horsemen training on a parade ground (*campus*). Campestres, no doubt, were first worshipped when soldiers began to train extensively on well laid-out parade grounds. This must have happened in Lower Germany where the army widely worshipped threefold goddesses – many, but not all, called Mothers. Since guards trained more than other units, and horsemen more than foot soldiers, the Campestres are likely to have first been worshipped at the training ground of the governor's *equites singulares* horse guard in Cologne. As the British Matres Ollototae, the 'Mothers of Other Peoples', became latinized as Matres Transmarinae, so the Campestres may have had at first a German name that became latinized when Trajan brought the horse guard to Rome. Weapons training on highly-strung horses was fraught with danger, hence the need was great for Campestres to safeguard the men as they exercised, whether for show or war.

Along the Rhine and the Danube, as in Britain and in Africa, horsemen worshipped the Campestres from Trajan's time on. In several places the dedicants of the altars were officers of the horse guard who may have brought the cult there. They changed the name of the goddesses to fit local custom: in Rome the Campestres, though feminine, are never called Matres, but in Britain they are twice so called. In Africa they are male *di Campestres*, in Spain Mars Campester. As guardsmen taught the training maneuvers sanctioned by the emperors to provincial cavalry units, they spread the cult of the Campestres in the hope of saving men from accidents. The horses themselves – even during exercises – stood in the care of Epona, for several times horsemen call upon Epona and the Campestres together.[186]

Suleviae appear thirteen times on the altars of the horse guard. Eleven times they are called Matres Suleviae, which seems to mean 'Suleviae Mothers' rather than 'Mothers *and* Suleviae'. Since the Mothers were worshipped mainly in the Lower German army, the guardsmen no doubt brought the cult of the Suleviae Mothers from Lower Germany to Rome. Their Celtic name means 'the well-leading ones', and indeed they are personal guardians 'who take care of you', for a guardsman set up an altar to the Suleviae Mothers of his father, his mother, and himself – a good son who cared about his aging parents far away on the banks of the Rhine.

C. Julius Caesar in his 'Gallic War' says of the Germans that they only worship gods they can see and who help them: the Sun, Fire (Vulcanus), and the Moon. Caesar wrote in a tradition that idealized the noble savage of the North. At the same time he slurred his enemies as primitive, and his account on the whole is wrong. Yet he also reflects the truth, for in 133 a guardsman of the Nemetes tribe (whom Caesar conquered in 57 BC) enlarged the usual list of gods by Sol, Luna, Terra, Caelus, Mare, and Neptunus (Rivers and Lakes) – all gods that can be seen.[187]

Oriental religions

Oriental gods, still shunned by guardsmen under Hadrian and Antoninus Pius, are first met under Marcus. The dedications of the horse guard thus constitute archaeological proof of the statement in the *Augustan History* that Hadrian scorned foreign religions while Marcus sought them out. They also show that the troops followed the lead of the emperors in their worship.

Three cult statues of Mithras have come to light around the forts on the Caelius. Hence three or more shrines of that cult flourished in the area, and though none is directly linked to the guard by inscriptions, guardsmen surely were among the faithful. One almost life-size statue of Pentelic marble (Plate 18) was found either inside the New Fort or just beyond its walls. Its size and quality point to wealthy worshippers such as guardsmen.

Despite its outward appearance, Mithraism was a Greek and Roman, not a Persian religion. Life springing from a slaughtered bull was a Greek and Roman vision. The elements of the cult icon are equatorial constellations: Canis Minor (dog), Hydra (snake), Scorpio (scorpion), Taurus (bull), and Orion (Mithras). Inspired by the invincible rise of the constellations midway through the heavens, Mithraism underpinned such Roman values as acceptance of one's role in society, endurance on journeys, and swift fulfillment of one's duties – all soldiers' virtues.

Constantine, it has been said, cashiered the praetorian guard because it turned a deaf ear to Christianity and held out for traditional Roman religion. The horse guard, too, has been called unchristian – and its overthrow the handiwork of providence. Recently found inscriptions, however, show Christians serving in the praetorian guard as early as the mid-third century, one of them even as a centurion. Very likely, therefore, some horsemen of the guard, too, were Christians by the third and fourth centuries. A few of their gravestones do not follow the pagan custom of calling on the spirits of the dead (*di Manes*). They may well have belonged to Christians. Two brothers in the *comites* detachments of the horse guard were clearly Christians.[188]

Honoring the emperor

From 128 onward, veterans of the horse guard often honored the *Genius imperatoris* – quite unlike other branches of the armed forces, who rarely mention the Genius of the emperor before the third century. For favors received, Romans thanked the Genius of the giver. The cult of the emperor's Genius thus seems to reflect the veteran horsemen's gratitude for cash awards upon discharge, just as in the Severan period the cult of the emperor's Genius by *beneficiarii* seems to reflect their gratitude for promotion.

From Caracalla onward, Roman army units very often broadcast their 'devotion to the emperor's godhead and majesty'. No such dedication is known from the horse guard, and there is only one inscription with the well-known formula 'If the emperor is safe, lucky is our unit'. This does not betray a lack of enthusiasm for the emperor, however, since so far only three dedications have been found from the time after Septimius Severus.

Nor is it surprising to see that only one of the horsemen's dedications, and none of their gravestones add an emperor's honorific title such as *Antoninianus* to the unit's name. Praetorian gravestones flaunt such titles, but those of the horse guard lack them because the phrase *domini nostri* in the unit's name proclaimed its closeness to the emperor even better.[189]

Gravestones

When, under Caligula, the *Germani corporis custodes* grew wealthy, they had huge gravestones carved for themselves, in a style borrowed from the praetorians: a wreath showed that the soul was deathless; spheres that it dwelt in heaven. The same beliefs were still held by guardsmen of the *equites singulares Augusti* in the second and third century, as may be seen from the wreaths carved on their gravestones and from Hadrian's poem for Zenodotus (p. 80). Yet when under

17 Altar to Silvanus, set up in the Old Fort by the veterans of 145. The god, his hair in the curly Antonine style, holds a cape with fruits and a pruning knife; his dog fawns on him.

Trajan the *equites singulares Augusti* came from Lower Germany to Rome, they brought along their own gravestone art. Their large, richly-carved stones nearly always show the same scenes: in a niche above the inscription, the horseman lies on a couch, dining, while below his groom brings up, on the long rein, a saddled horse (see Plates 2 and 14). Similar scenes can be found in the gravestone art of other provinces, but the long-reining of the horse by a groom, often with a coil of line in his hands, proves unmistakably that the guardsmen brought these images from Lower Germany.

Both scenes idealize the life of a horseman. The banquet, with its flower wreaths, food and drink, the service of a boy, and a luxuriously cushioned couch on carved legs, celebrates the good life. The richly bedecked steed, long-reined by the groom to warm it up, commemorates the noble status of a horseman. Such is the *imago genialis*, the ideal image in which the dead lives on.

Some gravestones show the 'Thracian Rider' scene (Plate 15), which, like the other two scenes, derives from Greek hero worship. Brought to Rome by horsemen from the lower Danube, it portrays a rider and his dog rushing a boar that faces them from a rocky cave at the foot of a tree. To hunt boars was to prove one's manliness, and some believed that to kill a fated boar, the *aper fatalis*, was to fulfill one's destiny. Emperors hunted boars as eagerly as the British in India went 'pig-sticking', and their guardsmen, of course, went along. The boar hunt on gravestones, however, stands for manliness in the abstract rather than for a duty of the horsemen. In one such scene a groom hands his master a helmet (see Plate 15), a heroic gesture, out of place on a hunt.[190]

On gravestones from Caracalla's reign onward, the dead often stand in full frontal view before their horses (see Plates 9, 10, 12, 14 and 16), looking at the viewer as enigmatically as men on Late Renaissance portraits. They are horsemen in their glory, though they boast not skill or strength but rather trappings of dress and equipment. For all their shortcomings, the reliefs give one the feeling that the horsemen stand there in person – the closest we come to bridge the gap of 1800 years.

18 Mithras killing the bull: from an underground sanctuary at the New Fort. Dog, snake, and scorpion join the god – all equatorial constellations breathing cosmic symbolism.

9
TRAINING FAITHFUL FRONTIER ARMIES

The foremost glory and the mainstay of the Roman empire,
steadfastly upheld sound and unimpaired to the present time, is
the firm bond of military discipline, in whose embrace and
keeping reposes our serene and tranquil state of blessed peace.

Valerius Maximus 2,7

Loyalty

'The fundamental problem of the principate', it has been said, ' – the problem which brought it into existence – was the control of the army'. One of Augustus' answers to that problem had been the guard, and the first official act of his principate was to double the guardsmen's pay. When Augustus died, the Pannonian legions rebelled, but the guard – the praetorians and the horse guard – quelled the mutiny. They were the last to forsake Nero, and in 218 in the battle at Antioch, the horse guard and the praetorians held out for Macrinus long after other troops had gone over to Elagabal. They were the most faithful of all units.

The loyalty of the horse guard, moreover, served to strengthen the faith of the frontier armies. Chosen from the *alae*, the guardsmen represented the frontier *auxilia* in Rome. They kept up bonds with their fellow soldiers on the faraway frontiers, even choosing them as heirs. From 192 on, when the praetorians were picked from the legions, they, too, spoke and acted for the frontier armies. It is not true, as has been said, that once in Rome, they turned their backs on their former units, despised the 'barbarian legions', or failed to help the group from which they had come. When Cassius Dio became consul for the second time in 229, Pannonian guardsmen hounded him from Rome for having ruled the Pannonian army with an iron fist.

146

It took some goading, therefore, to rouse frontier armies against the guard and the emperor. Would-be usurpers plied frontier soldiers with envy, telling them that guardsmen had it too good, what with high pay, privileges, and the city's theaters, baths, and circuses. Yet during the first 250 years of the empire, they succeeded only twice, once in 69 and again in 193.

A democratic trend in the army gave soldiers not only a role in bestowing decorations, meting out punishments, sending delegations to the emperor, or choosing under-officers and officers, but also in promoting men to the guard. Men so chosen would of course become spokesmen for their units. The guard, by letting the emperors know the wishes of the frontier armies, served as a link between the ruler and his warriors.[191]

Another way of strengthening the frontier armies' loyalty was to appoint guardsmen as their officers. We do not know whether some of the first-century *Germani corporis custodes* were farmed out as officers in the frontier armies (as were some *evocati* from the auxilia), for the need to bind the frontier armies to the emperor in this way grew strong only with the civil wars of 69. But of the second- and third-century *equites singulares Augusti*, many, after service in the guard, became decurions in provincial armies, where they served as troop leaders and drillmasters. As a rule, provincial governors themselves chose the decurions for their armies, yet men appointed by the emperor, understandably, took precedence. Proud of the emperor's favor, and flaunting the title *decurio factus ex numero equitum singularium Augusti*, they loomed as pillars of loyalty.

When commanding a field army, Trajan appointed even the lower-ranking *duplicarii*, for the loyalty of under-officers was not to be scoffed at. During the civil war in 69, for example, *ala Siliana* in Italy, in a decisive move, deserted Otho for Vitellius at the instigation of the decurions whom Vitellius, five years earlier as Proconsul of Africa, had 'obliged'. Provincial governors, who also had a horse guard of *singulares*, likewise tried to win backers in their armies by promoting guardsmen to decurions.

It is not known how many decurions of the *alae* came from the horse guard. One discharge list shows two out of 32 time-expired horsemen having become troop leaders in the provinces. Other men, no doubt, were promoted at a younger age. However many they were, their number was small compared to the 150 or so new decurions needed for the *alae* every year. Nevertheless, through their example (and perhaps also through reporting back to Rome), the emperor's men carried weight amongst the decurions of the *alae*.[192]

Winning the loyalty of legions through their centurions mattered even more. Although here, too, provincial army commanders had some freedom to choose, as a rule the emperors themselves appointed centurions. Such a promotion was a *beneficium*, a good turn that bound the recipient to the giver. Army records carefully noted the persons to whom centurions or decurions owed their appointment. An affair in the reign of Augustus highlights the bond that stemmed from promotion. When some centurions on furlough came to see Tiberius while he languished in exile on Rhodes in 6 BC–AD 2, their visit laid Tiberius open to charges of treason, since it was he who had raised the centurions to their rank (*beneficii sui centuriones*). In Roman eyes, any officer whom Tiberius had promoted must be his eager backer.

Appointing centurions thus was an emperor's best way to win faithful followers in the legions. Of the 120 centurions needed for the legions every year, most came from the legions themselves, but a fair number also came from the praetorians, and many, too, may have come from the horse guard where promotion was quick. Whether centurions were mostly of Roman-Italian or of provincial stock mattered less to an emperor than whether he could trust them. It was for their trustworthiness that horsemen of the guard became legionary centurions and for their loyalty that former drillmasters of the horse guard became chief centurions.[193]

Trustworthy centurions in the frontier armies were of the utmost importance to an emperor, for they could sway the armies. A telling instance of this is Britain's adherence to Vespasian in 69: the second Augustan legion there sided with Vespasian 22 years after he had been its commander, a span of time longer than the twenty years served by legionaries but well within the memory of long-serving centurions. Likewise, centurions chosen by Vitellius tried to keep Britain in his camp. Personal bonds held Roman society together, and the bonds created between emperors and frontier armies are the main reasons the military anarchy came as late as it did – the army stayed loyal even during the wild excesses of emperors like Commodus.[194]

Training the frontier horse

From Scotland to Arabia, Roman troops never had to cross an international border – such was the extent of the Roman empire and the majesty of the army that upheld it. Yet the very size of the empire meant that battle groups drafted from faraway provinces for a campaign often differed in their training and could not fight as a team. The first war fought by such detachments seems to have been Trajan's

Parthian war of 113–17, and the setbacks then suffered may have sharpened Hadrian's zeal for uniform, high training standards throughout. In 143, when Hadrian's new rules of training were in place, Aristides could boast:

> Men defend the frontier who will not think of
> flight and who are flawlessly joined to each
> other in the use of all weapons of war.

It fell to the guard – the praetorians and the horse guard – not only to devise and to keep up such training but also to bring it to the frontier armies. Since fighting techniques can be taught only by highly trained men, praetorian *evocati* and former horse guardsmen came to teach the legions and the *alae*.[195]

Little is known about cavalry training during the first century, but Arrian's treatise on tactics (above p. 112f) reflects the state of the art in 136. Augustus, Trajan, and Hadrian, each had passed decrees (*constitutiones*) on how to exercise the army, and these had been gathered in a handbook called *Disciplina*, or *Disciplina Augustorum*. In his treatise, Arrian gives an overview of this *Disciplina* insofar as it concerns cavalry training exercises. He stresses that Hadrian himself had made some of the rules more demanding: the emperor added another target on the standard course asked for exercises in the manner of Eastern bowmen and Danubian lancers.[196]

Whether they were his own inventions, devised by the frontier troops, or adopted from 'barbarians', Hadrian must have worked out these exercises with his horse guard of the *equites singulares Augusti* to see how they could be taught. No other unit had such skill and ethnic diversity, so many training officers, and such a variety of weapons; and no other unit (save, perhaps, the praetorian horse) trained together with the emperor. Arrian's treatise thus sets forth exercises of the emperor's horse guard for use by provincial horse guards and frontier *alae*.

Having such outstanding skills, some guardsmen who went as decurions to frontier armies are likely to have served as training officers. Others, as centurions, became interim commanders of auxiliary units and thus oversaw their training. Fittingly, as a war academy for the Roman cavalry, the horse guard worshipped more than any other unit the Campestres, the goddesses of the parade ground, and it is to these goddesses that altars in Rome record the thanks of newly promoted decurioins and centurions – no doubt because skill in training had won them their promotions.

During the third century, when officers no longer needed to be aristocrats, guardsmen could become outright commanders of auxiliary units, as did Celerinius Augendus, commander of *ala Pannoniorum* at Gemellae in Africa, and Sex. Iulius Iulianus, tribune of *numerus Syrorum* in Mauretania. Such men knew how to look after the training of their units. Promoting horsemen of the guard to field commanders worked so well it set a pattern even for the fourth-century army when the emperors' guardsmen (*protectores*) became tribunes of field units.

Nor was this the only way for the guard's training standards to reach the frontier armies. Some guardsmen became *exercitatores* of provincial horse guards, and a former such *exercitator*, Calventius Viator, escorted Hadrian to the training grounds of Africa to witness cavalry maneuvers there. Very likely Calventius did so not merely as *ad hoc* commander of the emperor's horse guard but as the emperor's adviser in training excercises. The horse guard decurion Iulius Maximinus, because of his exercising skills, became chief training officer of the imperial field army in Germany – and thence rose to the throne.[197]

Hadrian stressed training more than any other emperor, and he, it seems, introduced the worship of Disciplina. *Evocati* as training officers for the legions are known from his reign onward. Yet Trajan, too, exercised with the horse guard and wrote *constitutiones*. Indeed, Hadrian called him his *auctor* in matters of military discipline. By bringing back the *Batavi* horse guard, Trajan surely also meant to raise the training standards of the frontier cavalry.

Biased anti-Domitianic sources darkly hint that, if the training of the frontier armies was lax, bad emperors thought it all to the good. Any uprisings from that quarter would be less fearful. Domitian and his ilk 'were glad to see military excercises slip, to see minds and bodies flag, and swords get dull and rusty from neglect' – much like fifteenth-century Mamluk Sultans, accused of razing the Amirs' arrow-shooting ranges outside Cairo so as to lower the training standards of Amir regiments. Be that as it may, Trajan and Hadrian, with the help of the new *equites singulares Augusti*, set about to strengthen the battle-readiness of their frontier horse.

During the Early and High Empire the horse accounted for only one-tenth of the guard (aside from the praetorian horse, whose strength is unknown). The share of horse in the Roman army at large, and in the Chinese army during the same period, was likewise one-tenth. By the fourth century, however, the share of the horse in the Roman field army had grown to one-third, and the guard of the *scholae* had become all horse. The horsemen of the Eurasian steppes with their overmatching cavalry tactics had forced this change on

Rome. To face the threat from the steppe, the emperors enlarged their own cavalry force and, with the help of the horse guard, greatly raised its tactical skill. Uniform fighting standards soon allowed the massing of horse into vast cavalry armies, a tactic fully developed in 194, when such a force won the battle on Mt Amanus for Septimius Severus. Long based on infantry legions, the Roman army, during the third century, clearly became cavalry-based, a change in which the horse guard played a major role.

In short, guardsmen from the frontier armies, coming to Rome and returning again to the frontiers, kept the armies both loyal and up to their task. This through-flow, with its fast promotions, fulfilled the soldiers' wishes as well as the emperors' needs and hence was a true 'secret of empire'.[198]

10

DEATH AT THE MILVIAN BRIDGE

He fell in battle, fighting for the country.

Gravestone of a guardsman

Maxentius

Whenever the guard took to the field with the emperor, a force of *remansores* remained in Rome to look after the forts and the horsemen's families and to keep up the supply and training of horses. Indeed, the forts still belonged to the *equites singulares* in 312 when Constantine razed them. A dedication to Minerva set up by the *curatores* was still then in place, hence the building had been used by no one else, although the troopers came in full strength to Rome for the last time with Diocletian in 285 and in 303.

Ruling far from Rome, fearful of uprisings, and in need of troops, Diocletian greatly weakened the city's garrison. Though he did not reduce the number of praetorian cohorts or their strength, he took the bulk of the praetorians on his campaigns and kept them afterwards at his court in Nicomedia. Most *equites singulares* likewise went with the four emperors to their capitals. Galerius took his *comites* from Pannonia to Egypt and later to his new capital of Thessalonica. Few stayed behind in Rome.

In 306 when Galerius, then residing in Thessalonica, wanted to tax Rome, the people balked. When he also gave orders for the last soldiers to quit the forts of the city, the people and the soldiers alike turned to Maxentius, son of the retired emperor Maximian. Passed over in the succession to the throne earlier that year, Maxentius was living in a villa near the training ground of the horse guard on *via*

152

Labicana, and the troops who led the uprising and hailed him emperor seem to have been men of the horse guard. Described as *turmae praetoriae* they were perhaps praetorians, but more likely horse guardsmen, for *turma* was nearly always a cavalry unit. They stood under two officers, the commanders, it seems, of the two forts on the Caelian Hill. Further proof that horse guardsmen triggered the uprising comes from the fact that afterwards Constantine cashiered 'the praetorian legions and their *subsidia*, given to tyrannizing the city'. *Subsidia* could be elite troops, often horsemen, held in reserve behind the battle line for decisive action, the typical role of the horse guard. In this view the praetorians were no longer the guard but units of the line or 'legions', while the horse guard was the elite, the true guard.

The writers who describe these events do not name the horse guard directly, for it was less well known to their readers than the praetorians having changed its name several times. This explains the awkward phrases *turmae praetoriae* and *praetoriae legiones ac subsidia*. Had the praetorians staged the uprising, there would have been no point in mentioning the *subsidia*. The horse guard, therefore, may have taken the lead in proclaiming Maxentius emperor.[199]

In the eyes of his enemies, Maxentius bought the faithfulness of the guardsmen by squandering Rome's riches on them:

> This monster handed the treasures Rome had gathered from the whole world in 1060 years to those with whom he robbed the country. Granting to these butchers of their own country other men's wives, and the lives and goods of innocent people, he bound them to himself to the death.

In civil wars soldiers on both sides looked forward to hefty payoffs, those of the winners called 'generosity' (*liberalitas*), those of the losers 'ransacking' (*latrocinium*). The wealth of Maxentius' guardsmen may be seen in a gravestone set up, it seems, for one of his *equites singulares* near Brescia in northern Italy (see Plate 15). The inscription is lost, hence the date is not beyond question, but Maxentius had placed a keen cavalry force at Brescia to ward off inroads by Constantine's or Licinius' forces. The style of the gravestone also suggests Maxentius' reign, for the horseman's stern frontality mirrors Maxentius' portraits on coins, and heralds the icons of Byzantium.[200]

The battle at the Milvian Bridge

In 307 Galerius sent an army under his Caesar, Severus, from the East against the usurper Maxentius in Rome. Severus' soldiers, however, wanted to live in Rome, where they had been before. They soon switched sides and went over to Maxentius. Later, when Galerius himself came to Italy with a larger army, he fared no better. His troops, awed by the city's *maiestas*, felt it was wrong for Romans to war against Rome. Many of them, perhaps *comites*-guardsmen once stationed in Rome, forsook Galerius, who had to beat a hasty retreat.[201]

When Constantine crossed the Alps in 312, his mostly German army had no such qualms. The huge-limbed, eager fighters conquered northern Italy and bore down on Rome. Maxentius and his army met them in the open field north of the Milvian Bridge, upstream from Rome. There, on 28 October, six years to the day after they had hailed Maxentius emperor, the guardsmen fought their last battle.

The belief that he could flee across the river with his horse guard and take refuge in Rome may have influenced Maxentius' battlefield tactics. He placed his army with its back to the Tiber, hoping to take all thought of flight from the men. 'Generals . . . who cross rivers . . . in order that the soldiers may either stand and win, or if they turn to flight be killed, I am not wholly able to praise nor yet to blame', said the tactician Onasander. Many of Maxentius' horsemen of the line were Mauri, quick to attack and quick to flee, but between the enemy and the river, they had to make a stand. Much of the battle hinged on the horse, and the Mauri may have fought well, side by side with the praetorians. Alas, Maxentius' main force faltered without a fight, and treason opened Rome to the enemy.[202]

When his army gave way, Maxentius fled, and since the bridge was broken, he and the horse guard swam across the Tiber.

> Gulping down the wrongdoers, the Tiber swallowed
> [Maxentius], too, as he vainly tried with his steed
> and dazzling weapons to get away over
> the steep bank on the other side.

Coming from a speech given in Constantine's presence a year after the battle, these words are trustworthy since in such a setting basic facts were not likely to be misstated so soon afterwards.

Maxentius thus swam the whole width of the swollen river with his weapons. He could do so because he held on to his steed in the manner of a guardsman. His horse guard went with him, as shown in the relief on the arch of Constantine (Plate 20). Surely the emperor had trained for this feat together with the horse guard. Although they are wearing sleeved scale shirts – if one may trust the relief – they are not *clibanarii*, for *clibanarii* rode armored horses with whom one cannot swim. Scale

154

had long been worn by Roman horsemen, and during the third century had become the cuirass of choice, worn by emperors as well as their guards.

Having swam across with Maxentius, many troopers, like the emperor, may have found the south bank too steep where they reached it. When Constantine's horsemen came up, besetting them with spears and arrows, flight was cut off. They had to surrender or drown. Yet in swimming the river the guardsmen had shown their daring one last time. Faithful to the emperor to the last, they went under in a way that befitted their gallant history.[203]

Constantine's German *auxilia* won the battle because Maxentius' many thousands of Italians lacked the will to fight. Although the core troops stood their ground, those at the city walls caved in quickly. Long peace had made the inhabitants of Italy easy prey for invaders, while beyond the borders of the empire never-ending wars raised fierce fighters. For a time, the army's high standards in weaponry, training, logistics, discipline, and engineering, as well as its power to romanize recruits from the frontier areas, made up for that weakness – and the horse guard did much to uphold these standards. Arming and training the masses, on the other hand, was as dangerous and costly in times of peace as it was slow and uncertain in times of war. Hence the people of the empire grew unwarlike and foreigners came to serve as mercenaries. Reformers, aware of this and of its dire effects, went unheard. Now, the price was paid: when Maxentius' army yielded, Rome ceased to be a capital.[204]

Constantine's unwitting monument for the horse guard

In the aftermath of defeat, the guardsmen and their families were not massacred. The killing, we are told, ended with the fighting. Following the custom of the time, the survivors, shorn of their splendid dress and weapons, swelled the ranks of the winners. Sent north, they suffered the punishment of transfer to a lesser branch of service and became border patrols. A year after the battle a flatterer of Constantine gloated:

> Stripped of their wicked weapons and re-armed against foreign enemies, they have forgotten the delights of the Circus Maximus, of Pompey's Theater, and of the spas of the noblemen; they patrol the Rhine and the Danube, watching out for robbers.

Being highly trained elite troops, Maxentius' guardsmen may not have become border guards but rather the mailed (*catafractarii*) horse-men found in the cities of Gaul soon after 312.[205] Constantine needed to unburden his guilt for having made war on Rome, and the guard was his scapegoat. The horsemen who rode to the battle at the Milvian Bridge no doubt felt, like other Roman soldiers, that they were fighting for the rightful emperor and for country, *pro re publica*, as they had sworn they would. Constantine, however, gave out that they had ransacked Rome, that as a *factio* they had tyrannized the city, and that he had come as Rome's avenger.

Widespread treason by Maxentius' army during the battle at the Bridge suggests that there was some truth in the charge. For all his propaganda about Rome, Maxentius, had lost the hearts of the people because he allowed his guard to ride roughshod over civilians: he was said to have allowed the guard to slaughter people. The fault was not Maxentius' alone, it was typical of the time. Praetorianism barred the rulers from winning broad-based, popular support.

To punish the wrongdoers, Zosimus said, Constantine 'disbanded the praetorian soldiers and razed the forts in which they were stationed'. The plural 'forts' proves that the forts of the horse guard were wrecked together with the praetorian camp, a fact that archaeo-logists have rediscovered over the last hundred years. The horse guard, therefore, like the praetorians, was cashiered on 28 October, 312. Seeing this, another flatterer crowed:

> Rome has a new and lasting foundation,
> – all who might ruin her are done for,
> root and branch.

Constantine's own view of things is echoed by modern scholars who believe that the overthrow of the 'utterly unchristian', the *paganissimo* horse guard betrays the hand of 'divine providence visible'. Though some guardsmen were Christians, collectively they must have served as hunters, jailers, and executioners of Christians during the great persecution under Diocletian, for that was in their line of duty as it had been during the Pisonian conspiracy under Nero. Memory of it must have lingered, for while Maxentius was no persecutor, Christians never tired of hallowing their martyrs. Moreover, Constantine's troops bore the sign of Christ on their shields and the horse guard had fought them to the last. Very likely, therefore, Constantine's wrath against the guard had not only a political but also a religious motive. His razing of the New Fort, then, was symbolic, and so was placing Rome's principal church, the Basilica of St John Lateran, on top of the guard's shrine of the standards.[206]

Even more stunning is the symbolism in the fate of the horse guard's graveyard. As if laying low the forts were not enough, Constantine also tore up the graveyard in *comitatu*, an awesome punishment in ancient eyes. He gave orders to smash to pieces the hundreds of splendidly carved and inscribed marble headstones of the *equites singulares* and to build with the wreckage the foundation walls of a huge basilica, later called St Peter and Marcellinus. Only lackluster saints were buried nearby, yet there rose Constantine's own grave rotunda and one of Rome's most richly endowed churches.

The choice of the horse guard's graveyard as the site of Constantine's mausoleum gives pause. Many other sites were available closer to the city, and with greater saints at hand. Here, however, Constantine's heavy hand had fallen on the 'wicked', and here he wanted his triumph to last forever. His giant sarcophagus of polished porphyry, now in the Vatican, but once in the rotunda *in comitatu*, gleams with threatening horsemen riding around fettered prisoners of war. The bound, brawny captives on the ground, all wearing soldiers' belts, and some made out to be barbarians, seem to be Maxentius' guardsmen, tied, symbolically, forever at the place of their unit's overthrow. Constantine's mausoleum thus signifies his victory over Maxentius' horse guard, very likely with Christian overtones.

The horse guard caught more of Constantine's wrath than the praetorians, for during the third century the horse had become 'the queen of the battlefield'. In one of history's fair ironies, however, Constantine raised a striking monument to the horse guard. When he razed its forts and graveyard, he hoped thereby to blot out its memory – instead he saved it. The rubble of the forts and the graveyard safeguarded vast treasure troves of inscriptions and graven images. In the shrine of the Old Fort alone, 45 dedications were unearthed, more than in any other sanctuary of the Roman army; and the Basilica of St Peter and Marcellinus, built over the horse guard's graveyard, has yielded 609 headstones, more than all other known cavalry gravestones taken together.

When Frankish monks in the ninth century stole the saints' relics from the Basilica, the building was left to crumble. However, half of Constantine's mausoleum still stands as if its massive weight tried to stifle the gravestones beneath. The underground walls may hold as many gravestones still as have been unearthed so far, and whenever a new relief or inscription comes to light, Constantine's unwitting monument again brightens the history of the Roman Emperors' dauntless horse guard.[207]

CONCLUSION

The horse guard had strengthened the rule of the emperors, heightened the splendor of games and festivities in the capital, bound the frontier armies to Rome, and raised the war-readiness of the empire. All in all, it was remembered well. Suetonius called it tried and trusted (*multis experimentis fidelissima*), and when Dante saw horsemen thronging the emperor – on Trajan's Column, certainly – he thought of the justice. In return, Germans, and in the third century Illyrians as well, had shared in ruling the empire long before Illyrians became emperors, and long before Germanic nations took over the West. Caesar's genius thus had found a splendid use for Rome's northern neighbors, and it was a triumph of Roman statecraft to draw strength, for five hundred years, from the tall, fierce, and faithful men born on Rhine and Danube.

After 312, the horse guard grew larger still. Franks, Alamanni and Goths flooded to the *scholae* as troopers and officers-to-be. Roman officers and German warriors, they added to the splendor of Byzantium and became the nobility of medieval Europe.[208]

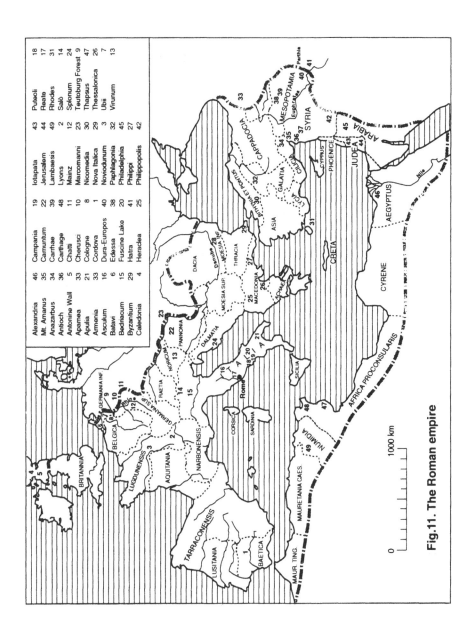

Alexandria	46	Campania	46	Iotapata	19	Puteoli	18
Mt. Amanus	35	Camuntum	35	Jerusalem	22	Reate	17
Anazarbus	34	Carrhae	34	Lambaesis	39	Rhodes	31
Antioch	36	Carthage	36	Lyons	48	Salò	14
Antonine Wall	5	Chatti	5	Mainz	11	Splonum	24
Apamea	33	Cherusci	33	Marcomanni	10	Teutoburg Forest	9
Apulia	21	Cologne	21	Nicomedia	8	Thapsus	47
Armenia	33	Cordova	33	Nova Italica	1	Thessalonica	26
Asculum	16	Dura-Europos	40	Noviodunum	40	Ubii	3
Batavi	6	Edessa	38	Paphlagonia	30	Vinunum	13
Bedriacum	15	Fuscine Lake	20	Philadelphia	32		
Byzantium	29	Hatra	41	Philippi	45		
Caledonia	4	Heraclea	25	Philippopolis	27		

Fig.11. The Roman empire

TIME CHART

Roman emperors' horse guard	58 BC –312 AD	Septimius Severus	193 –211
		Double guard of	193 –312
Germani corporis custodes	58 BC –68 AD	*equites singulares Augusti*	
Caesar	58 BC –44 BC	Caracalla	211 –217
Augustus	31 BC –14 AD	Macrinus	217 –218
Tiberius	14 –37	Elagabal	218 –222
Caligula	37 –41	Severus Alexander	222 –235
Claudius	41 –54	Maximinus	235 –238
Nero	54 –68	Gordian III	238 –244
Galba	68 –69	Philip	244 –249
Otho	69	Decius	249 –251
Vitellius	69	Valerian	251 –260
Flavian Emperors	69 –96	Gallienus	253 –268
Vespasian	69 –79	Aureolus	268
Titus	79 –81	Aurelian	270 –275
Domitian	81 –96	Probus	276 –282
Nerva	96 –98	The Later Roman Empire	284 –476
Trajan	98 –117		
Equites singulares Augusti	98 –312	Diocletian	284 –305
		Maximian	285 –310
Hadrian	117 –138	Galerius	293 –311
Antoninus Pius	138 –161	Constantine	306 –337
Marcus Aurelius	161 –180	Maxentius	306 –312
Lucius Verus	161 –169	End of the guard in Rome	312
Commodus	180 –192		
Pertinax	192 –193		

19 Epona, goddess of the horses: relief from the fort of Galerius' horse guard at Thessalonica. Sitting on a throne and wearing a wreath, the goddess feeds the horses.

GLOSSARY

ala	frontier cavalry regiment
auxilia	cohorts and *alae* raised in the provinces from non-citizens
Batavi	informal name for the bodyguard
Batavians	Germanic tribe in what is now the Netherlands
beneficium	a good turn, a boon, promotion
centurion	company commander; officer for 80 or more men
comites	'Companions', a late name for the bodyguard
corporis custodes	body guards, see Germani
decurion	troop leader, set over 30 men
duplicarius	cavalry under-officer with double pay
eques	horseman
equites singulares Augusti	'The emperor's own horsemen', name of the bodyguard 98–312
evocatus	a former praetorian soldier, re-enlisted
exercitator	drillmaster
gaesatus	fighter with a *gaesum* (an all-iron spear)
gentiles	tribesmen, foreign soldiers
Germani corporis custodes	'The German body guards', the emperors' bodyguard, 58 BC–AD 68
lanciarius	fighter with a *lancea* (javelin)
peregrini	soldiers of the frontier armies, doing special duty in Rome
pilum	lead-weighted spear

20 Battle at the Milvian Bridge in 312: relief from the Arch of Constantine in Rome. Constantine's bowmen and his horsemen in helmets and tunics drive Maxentius' scale-armored horse guard back into the Tiber.

praetorians	the emperors' foot guard
schola	club; under-officers' club; horse guard unit of the fourth century
sesquiplicarius	cavalry under-officer with pay and a half
singularis	guardsman
speculator	praetorian horseman of the emperor's escort
tribune	commander of a guard unit
vexillatio	detachment

FURTHER READING

For further reading about the Roman army and its guard see the following books, all but the second and fourth well illustrated.

M.C. Bishop and J.C.N. Coulston, *Roman Military Equipment from the Punic Wars to the Fall of Rome* (London 1993) well produced, with some very fine drawings.

J.B. Campbell, *The Emperor and the Roman Army 31 BC – AD 235* (Oxford 1984). The emperor-ideology, the rewards of service, the soldier and the law, the officers. Valuable, extensive use of literary sources.

P. Connolly, *Tiberius Claudius Maximus, The Legionary,* and *Tiberius Claudius Maximus, The Cavalryman* (both Oxford 1988). The career of a Roman soldier in superb artist's reconstructions.

R.I. Frank, *Scholae Palatinae. The Palace Guards of the Later Roman Empire* (Rome 1969). The horse guard of the fourth and fifth centuries.

R. Davies, *Service in the Roman Army* (New York 1989). The daily life, the duties, and service conditions in the army.

M. Durry, *Les cohortes prétoriennes* (Paris 1938). A masterpiece. Describes all aspects of the praetorian guard.

M. Grant, *The Roman Imperial Army* (New York 1974). A spirited account, mainly of the army in the empire's first century, with an eye on politics.

Y. Le Bohec, *The Imperial Roman Army* (Paris 1989, trans. London 1994). Wide-ranging, original, à good guide to the scholarly literature.

M.P. Speidel, *Die Denkmäler der Kaiserreiter* (Bonn 1993). A collection, with photographs and commentary, of the 776 stone monuments left by the *equites singulares Augusti*, the main body of information about the horse guard.

G.R. Watson, *The Roman Soldier* (London 1960). Written from the point of view of the individual serviceman and his career.

G. Webster, *The Roman Imperial Army* (3rd. edition, London 1985). The best available overall work on the Roman army in English, making good use of the archaeological sources.

NOTES

Abbreviations

AE	*L'Année Épigraphique*, Paris 1888ff.
ANRW	*Aufstieg und Niedergang der römischen Welt*, ed. H. Temporini, Berlin 1974ff.
ChLA	A. Bruckner & R. Marichal, *Chartae Latinae Antiquiores*, Zürich 1961ff.
CIL	*Corpus Inscriptionum Latinarum*, edited by Th. Mommsen *et al.*, Berlin 1872ff.
HA	*Historia Augusta*
RE	Pauly's *Realencyclopädie*, Leipzig 1893ff.
RIB	R.G. Collingwood & R.P. Wright, *The Roman Inscriptions of Britain*, I, Oxford 1965.
RIU	*Die römischen Inschriften Ungarns*, Budapest 1972ff.

Chapter 1 *From Caesar to Nero*

1. Foreword: Dante, *Purgatorio* 10,79f. Grosso in 1966, reviewing my 1965 dissertation, stressed the continuity from Augustus' *Germani corporis custodes* to Maxentius' horsemen and, as a consequence, their role as bodyguards throughout; his paper, borne out by the Anazarbos inscriptions (Sayar 1991 = Speidel 1993, 688-688e), and Bellen's work of 1984, are landmarks in research about the horse guard. Monuments of the *Germani corporis custodes*: below p. 25; monuments of the *equites singulares Augusti*: Speidel 1993. Noviodunum: Caesar, *BG* 7,13,1 *Germanos equites circiter CCCC summittit, quos ab initio habere secum instituerat*, see Bang 1906, 25. Full gallop of elite horsemen: Kromayer-Veith 1928, 370. The wording *secum habere* for escorts finds its parallels in Caesar, *BC* 1,75,2 and 2,40; also Suetonius, *Aug.* 49,1 and *Cal.* 43. Caesar's 400 men are mistaken for just another part of the army by Harmand 1967, 462, who was unaware of Bang 1906, 25; Durry 1938, 22: not a guard, but p. 75: a guard; Passerini 1939, 34: perhaps a guard; Bang 25ff and Bellen 1981, 48: auxiliaries; Millar 1977, 62: a guard.

2. Ariovistus: Caesar, *BG* 1,42. Foreign horse guards: Dessau 8888, see below, p. 124. Petreius (*BC* 2,75); Ptolemaeus (*BC* 3,4); Iuba of Mauretania (*BC* 2,40; see also 1,75,2); Herod (Josephus, *BI* 1, 397). Ubians: *Bellum Gallicum* 1,54; 2,35; 4,3; 4,8, and 4,11, also Bang 1906, 25; Bellen 1981, 36ff; and below, p. 40, fig. 2.

3. Spain: Caesar, *BC* 1,41, see Fröhlich 1882, 47; they, surely crossed the river *BC* 1,83. 800: Caesar, *BC* 3,106. Gauls: *B.Alex.* 17; Germans: *B.Alex* 29f; for the early Batavians see Willems 1984, 206–14; for cavalry tribes friendly to Caesar, Tausend 1988. Back to Rome: *B.Alex.* 77. Keen horseman: Suetonius, *Iul.* 57; Plutarch, *Caes.* 17,4. Instant readiness: Suetonius, *Iul.* 65. Travels, swimming: Suetonius, *Iul.* 57.

4. Labienus: Caesar, *B.Afr.* 19; compare Speidel 1993, 347; below, p. 66; 80. Truce: *B.Afr.* 29 – obviously Caesar's horsemen came from the same tribes (Bang 1906, 28). Body-strewn field: *B.Afr.* 40. *Fides*: Seneca, *De benef.* 5,34; *Beowulf* 2606; 2633ff; below p. 58 and 147f. Adamklissi, Metope inv. 32 (F.B. Florescu 36), see Strobel 1984, 235; Trajan's guard: below, p. 38ff. Batavians being huge: Tacitus, *Hist.* 5,18; *Ger.* 20. Strabo 8,290. *Commoda*: below, p. 94. Former slaves: Bellen 1981, 9ff. Caracalla, Galerius, see below, p. 66 and 75; compare Petreius' *familia*, Caesar, *BC* 1,75. Journey to Spain: Gelzer 1960, 273; compare Walker 1973, 342. Want of a guard: Caesar, *B.Hisp.* 2.

5. Cicero's letter: *Att.* 13,52. For the Spanish bodyguards see Suetonius, *Iul.* 86: *custodias Hispanorum cum gladiis <stip>antium se removisse*; also Suetonius, *Aug.* 49,1.

6. Dio 45,13,4; 46, 37,2; 47,48,2; see Bang 1906, 28f. Appian, *BC* 5,117 (485/6) as discussed by Bellen 1981, 15. Neither Bang (1906, 63 but see 28) nor Bellen link Caesar's guard with Augustus' *corporis custodes*, though the name of the unit, *Germani*, did not change. Trustworthiness: Tacitus, *Ger.* 28: *experimento fidei*.

7. *Batavi*: Suetonius, *Cal.* 43; inscriptions calling the *equites singulares Augusti* '*Batavi*': Speidel 1993, 688ff. Dio 55,24,6 calls Augustus' guard *Batavi* and picked auxiliary horsemen; he does not seem to have had the horse guard of his own time in mind (thus Mommsen 1881, 403; contra Bang 1906, 63f, Bellen 1981, and Speidel 1992, 112ff), for he says he cannot give their number which for his own time he would have known. Arrian's bodyguard were also horsemen, see below, n.33. For Batavians as elite soldiers see e.g. Strobel 1987; Speidel 1992. Picked: Speidel 1992, 112ff. Livy a Pompeian: Tacitus, *Ann.* 4, 34. Romulus: Livy 1, 15f; Dio 53, 16, 7; Syme 1939, 313f and 464. Peisistratos: Herodotos 1,59; 1,98. Pausanias: Thucydides 1, 130.

8. Sargon: Luckenbill 1927, 154 and 170. Persian king: Herodotos 7,40 (see also 9,63); Ptolemies and Seleucids: Geraci 1979. Tiridates of Armenia likewise had a horse guard of 1000 men: Tacitus, *Ann.* 13,38. Alexander: Tarn 1956, 11. Augustus: Suetonius, *Aug.* 25. Fort: Bellen 1981, 56f; praetorian horse: below, pp. 31ff.

9. Pusio: Dio 56,11, see Timpe 1970, 44f. If Dio had meant a regular auxiliary soldier, the term 'German' alone would have said it all; to add that Pusio was a horseman makes sense only if it was to show he was a member of the *Germani* horse guard (compare Dio 46, 37,2), very likely

Germanicus' own. For the body size and strength of the guardsmen see below, p. 79. Augustus' giant: Pliny, *NH* 7,74 and Reichert 1987, 543. Sieges: below, p. 123f. For Thracians throwing stones by hand against walls see Tacitus, *Ann.* 4,51; Aeimnestos: Plutarch, *Aristeides* 19; see Southern 1989, 109; guards throwing stones: below, p. 112f and 136. Germanic horse guards throwing stones: Caesar, *BG* 1,46; one pound: Vegetius 2,23. Stout guardsmen: Tacitus, *Ann.* 15,66.

10. Banished: Dio 56,23,4; Dio's term *apostellein* means to banish. The envoys may be the unarmed Gauls and Germans whom Augustus also banished. A parallel case is Suetonius, *Nero* 43: *conscios popularium, suorumque fautores*. Tribes West of the Rhine: Velleius 2,120, compare below, p. 29. For the Batavian custom of sending secret messengers from home to the troops see Tacitus, *Hist.* 4,15. Suetonius, *Aug.* 49,1 states they were cashiered, no doubt rightly so (contra: Bang 1906,64; Bellen 1981, 13 and 85), for being cashiered is a lighter punishment than exile. For their connection with the vanquishers of Varus see Walser 1951, 107 and Timpe 1970, 115.

11. In Rome in AD 14: Tacitus, *Ann.* 1,24. For guards of princes see Bellen 1981, 22ff. For the guard joining Tiberius in Illyricum in AD 6 see Velleius 2,111,3; every available soldier sent with Germanicus from Rome to Dalmatia Dio 55,31. For the role of the *Germani* as bodyguards see also Suetonius, *Aug.* 49,1. *Armigeri* served even private individuals as bodyguards, see *CIL* VI, 6229 and 9191 and Tacitus, *Ann.* 15,69. The oath of the guard, according to Epictetus 1,14,15 was to value nothing higher than the welfare of the emperor. Tyrant: Bellen 1981,83; fear: Millar 1977, 64; senator: Tacitus, *Ann.* 1,13,6.

12. Valerius Maximus 5,5,3; Pliny, *NH* 7,20,84; Bellen 1981, 28; Schumacher 1982, 6f. For Germanicus' horse guard see below, n.15. For Drusus' force see Tacitus, *Ann.* 1,24. Bellen 1981, 86f. For the use of the bodyguard against legionaries see also Caligula's move to Mainz, below, p. 22f, and Suetonius, *Galb.* 12; compare Keegan 1976, 203.

13. Picked horsemen: Tacitus, *Ann.* 2,16, see Speidel 1992, 114. Juvenal, 10, 95: *pila, cohortes, egregios equites, et castra domestica*, taken to be praetorian horsemen by Durry 1935, who was right, however, in seeing here soldiers of the guard where earlier interpreters saw equestrian noblemen. Dio 55,24,7, see above, p. 12.

14. Velleius Paterculus 2, 109: *Corpus suum custodientium imperium* (=discipline) *perpetuis exercitiis paene ad Romanae disciplinae formam redactum, brevi in eminens et nostro quoque imperio timendum perduxit fastigium*. Velleius was guard officer in the sense that he brought part of the guard from Rome to Illyricum (2,111). Since he had just finished his *militia equestris* and therefore had experience in commanding horse, he himself may have commanded the horse guard on that journey. For the guard in later centuries see below, p. 45ff; 109ff.

15. For Germanicus' and his son's guardsmen see Bellen 1981, 22-33, and above, p. 20. Not betraying their trust: the horse guard seems to be meant by Tacitus, *Ann.* 1,39 *praesidio auxiliarium equitum*; the *Batavi* are called auxiliary horsemen by Dio 55,24,6. Gifts: below, p. 25; Caesar: *B.Afr.* 40. German princes' *liberalitas*: Tacitus, *Ger.* 14. Puteoli: Suetonius, *Cal.* 19 and Dio 59,17,4; for such *Grossmanöver* see Horsmann 1991, 178ff. Planned war against Germany and Britain: Schumacher 1982, 13f. For Hadrian's guard see below, p. 45ff. For the emperor's role in battle see below, pp. 120f. Bridge, quality: Frontinus 4,2,1, compare Caesar, *BG* 4,17; Tacitus, *Ann.* 15,9; Pliny, *Ep.* 8,4; Dio 68,13.

16. For stories circulated see Suetonius, *Cal.* 43; also Dio 59, 21,2, see Bellen 1981, 34ff. Caligula at Valkenburg: Willems 1984, 229; for the horsemen hired see fig.2, below. p. 40. For the strength of the *Germani* see also Bellen 1981, 53-5. Josephus, *Ant.* 19,1,15 (122, see below, n.19), used the term *chiliarchos* advisedly, according to Grosso 1965b, 399f. Caligula, moreover, raised the number of praetorian cohorts from 9 to 12 (Tacitus, *Ann.* 4,5 and Dessau, 2701). Aurelius Victor 3,15: *externos barbarosque in exercitum cogere*.

17. For the maneuver see Suetonius, *Cal.* 45. Maneuvers: Balsdon 1934, 81; Barrett 1989, 134; Horsmann 1991, 178f. Swimming: below, p. 122f. False alarm by Trajan: Arrian, *Parth.* 41 (Roos); Dio 68,23; Caesar: Suetonius, *Iul.* 65. Trophies: Vergil, *Aen.* 11,5. Reconnaissance: Speidel 1992, 89ff and 1984, 183f. Keeping it up too long: Suetonius, *Cal.* 45. Ridicule: Dio 59, 25; Suetonius, *Cal.* 46; Tacitus, *Ger.* 37, 4 (*ludibrium*) and *Hist.* 4,15. Decimating the legions (at Köln, presumably): Suetonius, *Cal.* 48; for the horsemen there being guards see Bellen 1981, 88.

18. Thracians: Suetonius, *Cal.* 55; *archisomatophylax*: Philo, *Leg.* 27,176, see Grosso 1969, 227. Philo's words 'on the palace staff' equal the later term *domesticus*, they do not restrict Helicon to duties inside the palace; contra: Grosso, ibid.

19. Josephus: *Ant.* 19,1,15 (122), see Suetonius, *Cal.* 58; for the fury of German warriors see also Seneca, *De ira* 1,11,3f (below, notes 56 and 114). 'Uprising' Johannes Antiochenus fr. 84 M. = Dio 59,30,1b. Oath: below, p. 129; sense of duty: Tacitus, *Germ.* 14,1.

20. Batavians were *truces*: Lucanus, *Ph.* 1,431. Martialis, *Epigrams* 14,176. Mask: Brit. Mus. Blacas collection, GR 1867-5-8.644. For another mask of a Batavian (?), see Hänggi 1990. Red hair of the Batavians: Tacitus, *Hist.* 4, 61; Batavians dreaded in Rome in AD 70: ibid. 4, 68. For the hairknot see below, p. 130f. 'Foreign', see below, n.32. Purposely kept up: the Mauri in Martial's poem below, p. 41. Well-groomed: below, p. 104 and 130.

21. Sabinus' planned execution: Dio 60,28,2. Sabinus' rank as *chiliarchos* suggests he had been overall commander of the horse guard, not just commander of a guard detail; the same is true for Dio's term *stratopedar-*

chos, for which see Baillie-Reynolds 1923, 185; contra: Bellen 1981, 43f. For Claudius' freedman commander Actius (*CIL* VI, 4305) see below, n.23.

22. Gifts corrupting: Caesar, *B.Alex.* 48; Tacitus, *Hist.* 2,82. For the problem see Juvenal 6,347–8: *quis custodiet ipsos custodes?* Gravestones (25 have been found): Bellen 1981, 62f; 105ff. Giuliano 1984, 111–19 (praetorian stones 157ff); for a new monument of a *corporis custos* see Polverini 1982, 102f. Gamus: *AE* 1952, 147 = Bellen 1981, A, 14 = Priuli in Giuliano 1984, 117. Sphere: compare Speidel 1993, 18; beliefs: below, n.190.

23. Greek names: Bellen 1981, 76; *Gamus* is Greek (contra Bellen 1981, 74), see Reichert 1987, 306. Names on rolls: Mócsy 1992, 166-217. Suetonius, *Cal.* 55,2 and 58,3. *Speculator Augusti*: above, pp. 33ff; *frumentarius Augusti*: Clauss 1980. Germanus: Bellen 1981, 31. For the name *Batavi* see Suetonius, *Cal.* 43; Dio 55,24,7; Speidel 1993, 688–688b; manliest: Tacitus, *Ger.* 29. Curator: *CIL* VI 4305 = Dessau 1732 = Bellen 1981, A 24; also *CIL* VI, 20216 (Claudius Actius) – to understand with Bellen 1981, 66 *curator (collegii) Germanorum* is uncalled for. Compare the *curator statorum* (Dessau 2743) and the *curator veteranorum* (Domaszewski 1908, 79). Junior officers being *curator*: *CIL* III, 6025 = Dessau 2615, see Speidel 1992, 250. Two decurions are known to have been *curatores*: Laetus (*AE* 1968, 32 = Bellen A 7) and Spiculus (*CIL* VI, 8803 = Dessau 1730 = Bellen A 18 with Suetonius, *Nero* 30,2 and 47,3 and Dio 63,27,2b) see Bellen 43ff; for non-German decurions see also below p. 92. For the commander Sabinus see above, p. 25f. Collegia: Speidel 1993, 54, see also Suetonius, *Aug.* 32; Petrikovits 1975, 78ff and 132ff.

24. For Italicus see Tacitus, *Ann.* 2,9f and 11,16,f. For sons of soldiers serving in the guard see Speidel 1993, 30, for Italicus or Italus as names of horsemen see Speidel 1993, 114 and 713; fighting skills: below, p. 111. Swimming rivers: see below, p. 122; foreign noblemen in the guard: Frank 1969, 70f; Speidel 1993, 44 (the Parthian); below p. 80f.

25. Coin: BMC, *Nero* 143. For the imperial standard see p. 98 and 120. *Decursio*: Horsmann 1991, 183. Nero, as a boy, had borne a shield at the head of a praetorian *decursio* (Suetonius, *Nero* 7); the term *indicta*, 'ordered', suggests a military training maneuver rather than an honorific parade; stout commanders led such a *decursio campestris* personally, carrying a heavy shield (Suetonius, *Galba* 6,3). Emperors training with the guard: below, p. 115f.

26. Paramilitary: Bellen 1981, 49f. Seneca: *Clem.* 1,13,5. Gladiators as commanders and trainers: Suetonius, *Cal.* 55, see Bellen 43 and 67; to be sure, *Thraeces* gladiators fought on foot, but so, at times, did horse guardsmen (below, p. 123f); German horsemen fighting on foot: Caesar, *BG* 4,2. For gladiators as trainers of soldiers see Valerius Maximus 2,3,2; also Livy 44,9 and Vegetius 1,11; Le Roux 1990; Horsmann 1991, 55 and

135ff. Gladiators as a horse guard: Caesar, *BC* 1,14. Battle tactics taught by a *lanista*: Caesar, *B.Afr.* 71. Pliny: *Paneg.* 13,5 and below, pp. 42; 110. Slaughter: Dio 61,9,1.

27. Honor: Tacitus, *Ann.* 13,18. Compare Tacitus, *Ger.* 13 *in pace decus in bello praesidium.* Pisonian conspiracy: Tacitus, *Ann.* 15,58,2f. Vitellius' troops: Tacitus, *Hist.* 2,21;2,73, see Walser 1951, 70f. Parthian king: Tacitus, *Ann.* 6,36. Truly Roman bodyguards: Seneca, *Clem.* 1,13,2. Bellen 1981, 89 thought Nero thanked the *Germani* by putting them on coins, but see Speidel 1992, 115ff. Free grain: Tacitus, *Ann.* 15, 72; Suetonius, *Nero* 10, see *Iul.* 26. Gravestone with Ti. Claudii: *CIL* VI, 8803 = Dessau 1730 = Bellen 1981, no.18. Name and citizenship: below, p. 44; Bellen 67f prefers Germanic *Waffensohnschaft* but parallels with imperial names are lacking.

28. For Spiculus see *CIL* VI, 8803 = Dessau 1730 = Bellen 1981, 18; also Suetonius, *Nero* 30,2 and 47,3. Plutarch, *Galba* 8,6f; Dio 63,27,2b; Bellen 1981, 43ff; 66ff; 93f. Bellen 1981, 45 suggests Spiculus merely commanded the watch in the camp – if so, who was the commander of the horse guard? News of armies revolting: Suetonius, *Nero* 47; compare above, p. 18. Faithful unit: see p. 37.

29. For the end of Nero see Bellen, 1981, 93f. Bellen 1981, 97 suggests Galba cashiered the *Germani* because they had betrayed Nero, yet Suetonius (*Galba* 12) did not think so, nor Tacitus, *Hist.* 2,5. Back home, the former *Germani* may be meant by Tacitus' remark, *Hist.* 4,12 *erat et domi delectus eques*; Walser 1951, 89 sees in the *delectus eques* a *Landsturmtruppe*. Whether they are the Batavian *ala* of Tacitus, *Hist.* 4,18, or even the *ala Singularium* under the Batavian Briganticus (*Hist.* 4,70) remains an open question; see also Bellen 1981, 97ff and Speidel, 1992, 117. Willems 1984, 229 and Strobel 1987, 287 suggest that they later became part of *ala I Batavorum milliaria*.

30. Niggardliness: Suetonius, *Galba* 12. Suetonius: *Galba* 12; Caesar: *B.Afr.* 40. Law: Digest 49,16,3,21. Banishment from Rome: Digest 49,16,13; compare Dio 75,1 and Herodian 2,13,9. 'Strange': Suetonius, *Galba*, 20. For the moral state of the rest of the army at the time see Tacitus, *Hist.* 2,37 with Syme 1958, 185; also Otho's promise to the praetorians, Suetonius, *Otho* 6,3; also Aurelius Victor 26,6. Later all soldiers could expect to get rich: Tibullus 1,1; Vegetius 2,24. Bellen (1984, 97f) may be right in suggesting that Civilis commanded them on the way home.

31. Vespasian reducing the praetorian cohorts: Durry 1938, 80–1, see also *AE* 1978, 286; a Flavian emperor resenting the betrayal of Nero: Dio 67,14,4. Juvenal: 8,51f; Juvenal's themes are mainly from Domitian's time or older, see Syme 1984, 1135. Vespasian's *singulares*: Josephus, *BI* 3,97, see 3, 470. Grosso's (1965) arguments for a survival of the *corporis custodes* during the Flavian period are well met by Bellen 1981, 69ff, notwithstanding Reddé 1986, 526; see also *AE* 1988, 178.

32. For the praetorian horse see Domaszewski 1967, 23; Durry 1938, 99f; Passerini 1939, 69f. *CIL* VI 32638, used to calculate their strength, is not a reliable source: its 7 horsemen out of 68 praetorians may be too many, for the same list comprises also 3 *tubicines* – far too many for 68 soldiers. Hyginus 7, has 300 praetorian horsemen for 600 *equites singulares*, in 30 he has 400 praetorian horsemen for 450 *equites singulares*; proportions could vary for many reasons, hence one cannot say for certain that there were much more than 400 praetorian horsemen. If there were 100 horsemen per praetorian cohort, as in Hyginus 30, then their full strength at the time was 1000, but in expeditions a greater part of the horse took part than of the foot (Tacitus, *Ann.* 1,24). Legionary horsemen: Speidel 1994.

33. Quickly: *rapidi equis* (Tacitus, *Hist.* 1,40). Galba: Suetonius, *Gal.* 19, also Tacitus, *Hist.* 1,40. Nero: Dio 63,29. Panthers: Suetonius, *Claud.* 21. Battle on the lake: Tacitus, *Ann.* 12, 56; Suetonius *Claud.* 21; Pliny *NH* 33,19,63. Grant 1974, 138, suggests the horsemen served on foot, but showiness was the whole point. Crossbows for the horse guard: below, p. 105. Caligula: Suetonius, *Cal.* 45, see above, p. 22f. Competition: see below, n.39 and p. 60 and 73. Provincial field commanders: Arrian, *Ect.* 22f (*equites legionis* are meant). Cremona: Tacitus, *Hist.* 2,24f and 2,33.

34. Horsemen from the *alae*: Tacitus, *Hist.* 2,94. Staying on: Mommsen, 1910, 8 and Passerini 1938, 140f, contra Lieb 1986, 328; Durry 1938, 244 suggests only men of the 'good' provinces were retained, but Tacitus, *Hist.* 4,46 gives as the guiding principle for dismissals *culpa.* Auxiliaries married: Diploma 86 of Roxan 1985 (AD 113); marriage rights kept: *CIL* XVI, 25 (of AD 79 according to Nesselhauf followed by Passerini 140f, but of AD 71 according to Lieb, 1986, 324). A soldier of an auxiliary cohort as a praetorian under Vespasian: *AE* 1932,30, see Passerini 169.

35. For Hyginus see below, p. 45. Britain: *CIL* VI, 2464 = 32647 = Dessau 2089. Egypt: below, p. 73. Arruntius Claudianus: *AE* 1972, 572: *praef(ectus) ala[e ---] / vex[il]li prae[to]rianorum,* see Devijver 1976, A 166 and Strobel 1988, 271; 1989, 127; it is not known whether Arruntius led the *ala* and the *vexillum* at the same time; perhaps one should read *[trib(unus)] vex[il]li,* since the Greek text suggests he was tribune twice. Date and rank of battle awards: Strobel 1988, 271f; 1989; Maxfield 1981, 164 and 207. Already the commander of the *Germani* was comparable in some way to an urban tribune, see above, p. 22f.

36. For the praetorian *speculatores* see Clauss 1973, 46-79; also Domaszewski 1908, 20; Durry 1938, 108–10; Passerini 1939, 70–3, Lieb 1986, and Panciera in Giuliano 1984, 158ff. Lance: *AE* 1955, 24, see Bonanome in Giuliano 1984, 198; Clauss 79; add *AE* 1989, 134. Galba nearly wounded: Suetonius, *Galba* 18. Clauss 1973, 78 assumes a real fight, a *Handgemenge*; Grant 1974, 91, apparently thought the emperor alighted into the lance of the *speculator*, for he sees him on foot and standing very close by. Spear butts: below, p. 130f. Sharp weapons: Tacitus, *Ann.* 14, 61 hinting at Nero's tyrannical ways. Galba cut down by horsemen: Suetonius, *Galba*

19; likewise Nero: Dio 63,29. The *lancea* of the *speculatores* with its long shaft differs from the much shorter weapons of the later *lancearii* (Balty 1987, fig.5). Spearblades sheathed (*praepilatae*): Caesar, *B.Afr.* 72.

37. *Speculatores* strength: Clauss 49ff; *Celeres*: Livy 1,15,8; praetorian horsemen: Kromayer-Veith 1928, 311; *excubitores*: Frank 1969, 206ff. Sparta: Herodotos 7, 205 and 8, 124. For Napoleon's 300 *gendarmes d'élite* see below, p. 131. Bedriacum: Tacitus, *Hist.* 2,33. Escort: Tacitus, *Hist.* 2,11. Drillmaster: M. Vettius Valens, *CIL* XI, 395 = Dessau 2648, for which see Clauss 1973a, 55 and below, pp. 110f. Battle decorations: *AE* 1954, 162.

38. Thus Clauss 1973, 47f. Three newly found gravestones of the later first century (*AE* 1976, 18; 21; 22) bear this out in that they do not add *Augusti* or *Caesaris*.

39. State banquets: Suetonius, *Cl.* 35; Dio 60,3. Tacitus: *Ann.* 15,58,2 *quibus fidebat princeps quasi externis*. For *externus* meaning barbarian see Walser 1951, 67–72. Greek tyrants characteristically had bodyguards of savage barbarians: Cicero, *Tusc.* 5,58; see Bellen 1981, 83. Grant 1974, 91 suggests that Augustus felt safer when protected by two rival corps rather than by one. Galba: Suetonius, *Galba* 12. Older men: Clauss 1973a, 58, for an exception see Panciera in Giuliano 1984, 162f: only six years of service; sensitive tasks: see below, p. 63.

40. Otho, Flavians: Tacitus, *Hist.* 1, 21-41; Suetonius, *Galba* 19; Clauss 50; Durry 1938, 372-8. Diplomas: Roxan 1978, no. 1 and *CIL* XVI, 21, see Lieb 1986, 324. *Speculatores* as bodyguards no longer heard of after 68: Clauss, 1973, 57; merged with the praetorians before 136: Lieb 1986, 326; for their being part of the praetorium even earlier, see S. Panciera in Giuliano 1984, 162f. *Singulares* replacing *speculatores*: Hirschfeld 1913, 587f, though there is no reason to assume the *speculatores* lost their security police function. Hyginus being Trajanic: Lenoir 1979, 111ff and Strobel 1984, 105. For *hastiliarii* see below, pp. 43ff and 130f.

41. Provincial guards: Speidel 1978. Titus' fighting prowess: Suetonius, *Titus* 4,3; his *singulares*: Josephus, *BI* 5,47; picked from the *alae*: Josephus, *BI* 3,97; six hundred: 5,52; since Augustus' time: Speidel 1978, 5; reconnoitering: Josephus, *BI* 5,52, see also 5,258 and below, p. 119. Josephus as commander of the Galilean forces organized along Roman lines likewise had 600 *singulares* 'as bodyguards' (*BI* 2,583). Saving the tenth legion: 5,82–97; siege works: 5,288 and 5,486. Archers: 5,331–341, compare Suetonius, *Titus* 5 and see below, p. 105; compare 6,70 where Titus also entered and fought with *singulares*; last attack: 6,246. They begged: Suetonius, *Titus* 5. *Ala I Flavia praetoria singularium* known from diplomas in Syria in AD 88 and later: Roxan 1978, 3 and 4. For the origin of *alae* from *equites singulares* see Speidel 1978, 54ff; I was wrong, 1978, 62, in thinking the *ala* was the same as the like-named one in Pannonia, see Strobel 1984, 116.

42. Caligula: above, p. 22. An argument for *equites singulares Augusti* under Domitian is made by Pitts-St Joseph 1985, 168, though *CIL* VI 3255 is no proof, see Speidel 1993, 672. M. Ulpius Singularis, recruited in 104 (Speidel 1993, 3) and thus born in 84–7 could owe his *cognomen* to his father's service with the guard, perhaps Domitian's *equites singulares Augusti*. For Domitian's campaigns see Strobel 1989, for his stay in Mainz also Walser 1989. Plate 1: Speidel 1993, 684; photograph Landesmuseum Mainz. Eastern horsemen: *CIL* III, 13483a = Dessau 9168 (Carnuntum), see Strobel 1989, 20ff. Battle in the Circus: Suetonius, *Dom.* 4 and Dio 67,8. Greek: Pliny, *Paneg.* 13, see below, pp. 110f.

43. Wary eye: see below, p. 67. Domitian's murder: Suetonius, *Dom.* 16ff; Dio 67, 14ff. Grosso 1964, 228 suggests that Parthenius, one of the plotters, was the commander of the bodyguard (since he had a right to bear a sword). Suetonius: *Galba* 43.

Chapter 2 *Riding High: The Second Century*

44. For the beginning of Trajan's reign see Dio 68,3ff and Syme 1958, 10ff; for his command of the Lower German army the military diploma of February 20, AD 98. For *singulares* of the governors see Speidel 1978. One thousand: Hyginus 8. Preferred: Cheesman 1914, 26; Lenoir 1979, 126f. *Singularis* = "own": Cicero, *De re pub.* 1,33 (50) and 2,9 (15); diplomas: Speidel 1993, 76 and 79. I was wrong about *singularis* in 1978, 4. When the unit was still new, some called themselves *singularis imperatoris Traiani*: Speidel 1993, 20 and 21. *Batavi*-name: Speidel 1993, 688–688e; also 75a and 757; below, p. 62.

45. History: Suetonius, *Galba* 12; Dio 55,24,6; Suetonius did not believe Cluvius Rufus' story of their 'betraying' Nero (for the tangled sources of these opinions see Grosso 1965,b, 403ff). Honor of the *Batavi*: Tacitus, *Ger.* 29. Finest horsemen: Dio 55,24; Tacitus, ibid. The special status of the Batavians was perhaps abolished after their uprising, see Syme 1958, 127; Tacitus, *Hist.* 5,25 and Strobel 1987, 286. Reviled: Juvenal 8,51; see above, p. 30.

46. *Corporis custodes*: Bellen 1981, 107–13; *equites singulares Augusti*: Speidel 1993, 110; 112; 136; 144; 166; 173; 181; 211; 277; 284 (Batavi); 2, 82, 170; 175; 204; 298?; 671; 728 (Claudia Ara); 3 (3 men, but counted as one piece of evidence); 341 (Ulpia Traiana); 245; 346 (Canninefates); 137; 174; 275 (Marsaci); 101; 103; 159; 180; 202 (Frisiavones, including *CIL* VI, 3220 taken for a Frisius by Will 1987, 30); Nero's Frisiao: *CIL* VI, 4343 = Dessau 1721. For Caligula and the Canninefates see Tacitus, *Hist.* 4,15,2; Will 1987, 22. For the discussion about the Tungrians (but not Frisiavones) belonging to Belgica see Bogaers 1972; Will 1987, 24f; also below, p. 54f.

47. *Comitatus*, Batavians: Tacitus, *Ger.* 13–14; 29; *Hist.* 4,12; see Syme 1958, 129; Timpe 1987; Lund 1991, 2073–8. Oath of the guard: Epictetus

1,14,15. Why Tacitus wrote the *Germania*: Lund 1991, 1954–6; *fides*: above, p. 14; ascribing deeds: Tacitus, *Agr.* 8,3; Josephus, *BI* 3,298, see Lund 1991, 1899f. Suetonius: *Galba* 12. Greek names: A. Birley 1993, 55. Trajan inspected the armies: Dessau 1019. Fighting skills: e.g. *contarii* (below, pp. 111ff). Choice of gods: below, pp. 141ff. Gravestone motifs: Speidel 1993, 4–8, and below, p. 144f. Plate 2: Speidel 1993, 205. Photograph: Musei Vaticani XXXIV–35–19–1. Documentary sources: Speidel 1993, 72.

48. Martialis 10,6. Mauri: perhaps *cursores et Numidae* (with their own *exercitator*) see Dessau 1714–16; for Mauri in the horse guard: below, p. 84. Trajan's Column, scene LXXXIX. Without bodyguard: Dio 68,7,3 and 68,15,5, see 75,15,4. Constantine: Eusebius, *VC* 3,10,2f; 1,44,2. For Trajan's soldiers kept well in hand see Pliny, *Paneg.* 22 and 23; Dio 68,7,5 and 75,2,2–6; for Hadrian's soldiers see Dio 69,5,2. Severus: below, p. 57. Emperor traveling: Pliny, *Paneg.* 20 says that Trajan had a *comitatus accinctus et parens*, a fine staff for such things, very different, for example from Caracalla (Dio 78,9) or Valerian (Zosimus 1,36). Caesar: above, p. 15. Guard in towns: below, p. 62. Legionary horsemen: Speidel 1994. Auxiliaries fighting without spilling Roman blood: Tacitus, *Agr.* 35. Training officers: below, p. 110ff.

49. Fort: below, p. 126ff; gravestones, p. 144f. Hastiliarii in AD 135: Speidel 1993, 7 and below, p. 130f. Trajan putting praetorians out of the way: Dio 68,5. Trajan's reaction to the praetorian mutiny: Durry 1983, 80f and 378f. For *speculatores* see above, p. 33ff; for *hastiliarii* see below, p. 130f; a good parallel is Germanicus' use of auxiliary horsemen in AD 14 when the legions rioted: Tacitus, *Ann.* 1, 39. Kienast 1966, 72–5 makes a case for the fleets being used by Domitian as a counterweight to the praetorians; see likewise Herodian 8,2. Septimius Severus: below, p. 57f. New guard units in history: Bartusis 1992.

50. For the *corporis custodes*' command structure being semi-military see Bellen 1981, 46ff. Suetonius, *Cal.* 43: *numerus Batavorum*; for the *numeri* of the provincial *singulares* see Speidel 1978, 22ff. The term *numerus* had no technical meaning, see Speidel 1984, 119 and 199; Southern 1989, 83ff.

51. *Singulares Augusti*: Speidel 1993, 72 (but see ibid. 20 and 21). *Imperator*: Nesselhauf, 1937. Tradition: Speidel 1993, 684. Name of the unit: ibid. p. 14; see Le Bohec 1989, 566.

52. Reliefs: Speidel 1993, 416f. Trajan's brave deeds and soldiers running risks for him: Dio 68,14. Soldiers' names etc. see Pliny, *Paneg.* 15; also Fronto, *Princ. Hist.* 9.

53. *Hastiliarii*: below, p. 130f: Trajan's Column, scenes LXXXIX and XCVII.

54. Witness deeds: Josephus 6,135. Hatra: Dio 68,31 (compare Septimius Severus' guards shot by the Hatreni, Dio 75,11 and below, 62); for the siege works at Hatra see Kennedy-Riley 1990, 105ff; for horse guards in

sieges below, p. 123f. Sharing the risk: Sallust, *Cat.* 59,1; Caesar, *BG* 1,25; Tacitus, *Agr.* 18. Hyginus, *De mun. cast.* 7 and 29, see Strobel 1984, 105f; Vegetius 3,23 says camels are useless in battle: surely in the West?

55. *CIL* III, 3676, see page 1042 = Dessau 2558 = Bücheler 1887 427, see Speidel 1991. The translation borrows from Davies 1989, 111. Written by Hadrian: see below, p. 80.

56. Dio 69,9; even if the horseman belonged to the local *ala I Batavorum milliaria*, known from a new diploma in Pannonia in 112, the tactics and strategy were the same in the horse guard. German, Parthian horse guards: Tacitus, *Ann.* 2,9; 15,15. River crossing and archers: below, p. 122f. Soranus is a Syrian name, see RIU 586. Defense value of rivers: Nesselhauf 1952, 234f. German pluck and Roman discipline: Seneca, *De ira* 1,1,3f; Herodian 8,1,3. Strategic value of maneuvers: Caligula at Puteoli, above, p. 21f. Marcomannic guard: above, p. 20. Skill and bravery, not numbers, win wars: Josephus, *BI* 3, 478; Vegetius 3,1. Ammianus 25,6,13 – 25,7,3 and 25,8,1, see Speidel 1991. Tacitus: *Ger.* 13. See also the close parallel (*topos?*) Tacitus, *Agr.* 18. On German leaders' followers see Lund 1991, 1899.

57. Altar: *CIL* III 1904 = Dessau 2417 = *IDR* III/2, 205 (this drawing). Campestres: E. Birley 1988, 419ff; 433ff; Speidel 1992, 290–7. A further dedication to the Campestres and Epona from Pföhring/Raetia (*CIL* III 5910 = 11909 = Dessau 4836 = Vollmer 261) was set up by Aelius Bassianus, perhaps also formerly of the horse guard. A parallel to Viator's career is that of T. Aurelius Decimus (*CIL* II, 4038 = Dessau 2416 = Alföldy 1975 no. 38) who became trainer and commander of the Spanish *equites singulares* in Tarraco, see E. Birley 1988, 404 and 421. For transfers to frontier armies see below, p. 147ff.

58. On the plot of the four ex-consuls see Dio 69,2,5; *HA, Hadr.* 7; Syme 1958, 244f. and 599f; 1984, 1281. Calventius Viator: Leglay 1974; Speidel 1978a, 29; Speidel 1993, 22. The Sarmizegethusa inscription: *CIL* III, 1904 = Dessau 2427 = Speidel 1978a 31; for the pluralization of Epona see Rüger 1987, 3, for the dative plural in *-abus* Schmitt 1987, 147. A parallel for a provincial *exercitator*'s promotion to the Rome guard is *CIL* VI, 533 = Dessau 2088. For Hadrian's *frumentarii*-snoops see *HA, Hadr.* 11,4.

59. Lambaesis speech: Dessau 9134; Leglay 1974; A. Birley warns me that Calventius' identification with the Viator of Lambaesis is not certain. Gerasa: Speidel 1993, 22. Rewards for other informers on high treason: Tacitus, *Hist.* 2,75; Suetonius, *Otho* 2; Dio 74,32,4; 74,10; Ammianus Marcellinus 18,3,6; Codex Theodosianus 7,1,10; and the career of M. Aquilius Felix, Dobson 1978, 278. Tribune known under Hadrian: Speidel 1993, 73 and 74. For the *hastiliarii* see below, p. 130f. *A pugione:* see below, p. 52; Grosso 1964, 227. Regulations: below, p. 112f. Soldiers hating to travel: Tacitus, *Hist.*, 1, 24. Hadrian's travels: Halfmann 1986, 188ff (Gerasa p. 206; Herakleia p. 203); Herakleia gravestone: Speidel 1993, 686.

60. Cocceius Firmus and other such promotions: below, p. 140. For Pius' funeral on the Antonine Column see Speidel 1993, 418. Medallion: BMC Med. Marcus 12. *Profectio*: below, p. 97.

61. Plate 3: Aurelian Column, scene LXXIV; photograph Deutsches Archäologisches Institut, Rome, Inst. Neg. 89,359. Parades: Arrian, *Tact.* 36,1; Josephus, *BI* 5,350. *Hastiliarii*: below p. 130ff; no blades needed: Speidel 1992a, 24–6. Carnuntum: Speidel 1993, 685. For the setting of gravestones see below, p. 62.

62. *HA, Com.* 12,7: *datus in perpetuum ab exercitu et senatu in domo Palatina Commodiana conservandus*, see Speidel, 'Commodus', forthcoming. Compare *HA, Com.* 2: *commendatus militibus*. Sign of strength: Halfmann 1986, 45ff.

63. Quintianus: Dio 72,4,4; Herodian 1,8,6. Perhaps Herodian merely expanded on Dio; nevertheless, he knew how the imperial court worked, for he himself had been a courtier (Grosso 1964, 33f; Whittaker 1969, 76). Guard expanded: Herodian 1,11,5. *A pugione*: Herodian 1, 12, 3; *HA, Comm.* 6,12–13; *AE* 1952, 6 and *AE* 1961, 280, see Grosso 1964, 217ff. If Cleander replaced the tribunes, his name must be supplied in Speidel, 1993, 19. Other commanders: below, p. 100. Cleander's end: Dio 72,13. Herodian's account of the role of the guard in the fall of Perennis is an invention, see Speidel 1993, 413. *Invisior Cleandro*: Ammianus 26,6,8.

64. Tyrant: see also Caracalla, below, pp. 65f. Cleander, Herodian 1,12,3 and *AE* 1952,6. Residence on the Caelius: *HA, Comm.* 16; nightsticks and axes: Herodian 2,4,1; 2,6,10, see Speidel 1993b and Lactantius, *De mort.* 12,5. Jealousy: Rostovtzeff 1957, 399, see Speidel 1993a. Lurius Lucullus: Rostovtzeff 1957, 398 with Dessau 6870 (doubted by Grosso 1964, 138). Newly-found gravestone: Speidel 1993, 219 (if, as is possible, the last name starts with a letter other than L, Lucullus may have been a relative of the squadron leader). For similar petitions and soldiers see Rostovtzeff 1957, 478; Durry 1938, 284ff; Millar 1977, 246; Campbell 1984, 264ff; MAMA X, 114; Turpin JRS 1991. Caesar and Augustus both personally helped their soldier-followers, see Seneca, *De benef.* 5,34 and Suetonius, *Aug.* 56. His man: Gilliam 1986, 304; Speidel 1978a, 67 and 1992, 81.

65. Commodus, his teachers and his hunting: Herodian 1,15,1ff; Whittaker 1969, 101. Compare Domitian, above, p. 37. A Parthian among the horse guard: Speidel 1993, 44; a Maurus as training officer of the horse guard: Speidel 1993, 755. Emperors training with the guard: below, p. 115f. No Moorish unit in Rome: Lieb 1986, 339 contra Domaszewski 1967, 137, 164 and 189, Durry 1938, 36, and E. Birley 1988, 23. Liked by soldiers: Speidel 1993a.

66. The guard going soft: Dio 74,16; Herodian 2,10,6; *HA, Did.Iul.* 5,9, see below, p. 132. Barbarian legions: *CIL* V, 923 = Dessau 2671 (second century, contra Durry 1938, 250). *Militiae provinciales* less trustworthy: Hyginus, *De Mun. Cast.* 2. Wishful thinking: *Paneg. Lat.* 12,14,6.

67. For Pertinax' murder see the *HA, Pert.* 11: *Tausius quidam, unus e Tungris*; Dio 74,9. Tausius was thought to be a *singularis* by Bang 1906, 86; Durry 1938, 382, n.4; Grosso 1966, 149; Whittaker 1969, 169; Smeesters 1977, 184. Second-century praetorians from Lower Germany: Speidel 1993, 744 with literature. Tungrians as praetorians: *CIL* VI,32623, 28 (with the name of the tribe, not the town of Atuatuca, just as in the *HA*) and *CIL* III, 5450 (dated to the second century by Weber 1969, 47), see also Smeesters 1977. The Tungrians belonged perhaps to Belgica, see above, p. 39. For the horsemen's detachment in the palace see Dio 74,9; for such detachments in the first century see above, p. 23. Grosso 1966 interprets Tertullian, *Apol.* 35,8–10 *qui inter duas laurus obsident Caesarem* to be guardsmen who killed Pertinax, but the expression *inter duas laurus* was symbolic, not a place name, hence not the name of the horse guard, see Seeliger 1987, 72.

Chapter 3 *The Roughshod Third Century*

68. Altars, gravestones: Speidel 1993, 63; 64; 578; 580; see also 75a. No more gravestones: MacMullen 1982.

69. Tacitus: *Hist.* 2,75. Severus: Dio 74,15; cavalry dress: Dio 75,3. Titus etc: Josephus, *BI* 3,470; 5,52; compare 2,583. 600 on a mission: Josephus, *BI* 4,57. *Ala* of 600: Lydus, *De mag.* 1,46 (for what it is worth).

70. Severus' entry: Dio 75,1ff; *HA, Sev.* 7; celebrate: Speidel 1993, 55 and 58. Cashiered, shocked: Dio 75,2. The new guard: Dio 75,1ff; *beneficium* obligation: Seneca, *De benef.* and *De clem.* 1,13,5; Garnsey-Saller 1987, 148ff; below p. 147f; MacMullen 1986; enrollment as *beneficium*: Speidel 1992, 306ff; those promoted to the horse guard were *beneficiarii*: Caesar, *BC* 2,75. Career promotion, of course, was also such a *beneficium*, see below, p. 96f. Spokesmen: Dio 80,4,2. Chosen: Speidel 1992, 158, see below, p. 146f. Lack of principles: Campbell 1984, 117; as for loyalty, the praetorians avenged the murders of Domitian and Commodus (Suetonius, *Dom.* 23; Dio 68,5; compare Herodian 2,6,10; *HA, Did.Iul.* 2,6).

71. Loyalty to the whole house: Tacitus, *Ann.* 14,7; below p. 102. Funeral: Dio 75,4 and 5; compare Josephus, *BI* 1,672 for the guard's role in the funeral of Herod. New Fort: below, p. 128ff. Minerva: Speidel 1993, 54. *Scholae*: below, p. 75. Inscriptions: Speidel 1993, 57; 60. Ranking in frontier armies: *ChLA* VII, p. 2. Commanders of 1000 horse always outranked those of 500 horse, see E. Birley, 1988, 349–64.

72. Gravestones mentioning the *castra*: Speidel 1993, 14; 25; 95. Not mentioning it: Speidel 1993, 524; 530; 541; 542; 553; 554; 555; 558; 559; etc. One *numerus*: 55; 60. Common dedication: 69. One Genius: 57; 59 (Genii show with which units the soldiers identified themselves, see Speidel 1984, 355–7). Diplomas: 76–9. Old-style reliefs on new-fort gravestones: 635; 657. *Batavi*: 688; 688c; 688d; see below, p. 62. *Lions*:

below, p. 64ff. Outdo one another: Tacitus, *Hist.* 3,27 & 29; Josephus, *BI* 3,485; Vegetius 1,23 and 3,4; Speidel 1992a, 32; above, p. 32 and n.39.

73. Doubling: Durry 1938, 88; the praetorians, even before, were 10,000 strong, see Kennedy 1978. See also Rostovtzeff 1957, 402; Campbell 1984, 401ff; and, for the field army, E. Birley, 1984, 21ff. Ruffians etc: Dio 75,2. Capital: below, p. 155.

74. Alexander: above, p. 17. Lyon: Dio 75,6; Herodian 3,7,3; *HA, Sev.* 11, compare Suetonius, *Aug.* 10; A. Birley 1988, 125. British: Herodian 3,7,2. Galba's coin: *RIC* 464 (Bellen 1981, fig. 22). For the setting of *adlocutiones* see Campbell 1984, 69ff. Fig. 4: drawing by P.S. Bartoli, after Brilliant 1967. Plate 5: panel IV of Severus' Arch on the Forum; photograph: Fototeca Unione at the American Academy in Rome, Askew Collection, plates 5 (SW 13) and 9 (SW 6). Third-century gravestones: plates 7; 12; 15; 16. Hatra: Dio 75,11,3; two bolts: see Ames 1993, fig 11. Trajan: above, p. 45.

75. Anazarbos: Speidel 1993, 688–688e; Sayar 1991. Gerasa: Speidel 1993, 22. Apamea: Speidel 1993, 689; 690. Honor: K. Ziegler 1985, 76; 137; 141. Boons of the emperor's (or his guard's) presence: Halfmann, 1986, 111–42. See also Trajan's, Caracalla's and Valerianus' staffs above, p. 42. Burden: Rostovtzeff 1933, 302–4. *Batavi*: see above, p. 39. Gravestones abroad: below, p. 117f.

76. *Curatores*: Speidel 1993, 58; A. Birley 1988, 146ff. Bulla: The story is Dio's (76,10); for comparable use of regular cavalry against brigands see Davies 1989, 154. Plautianus: Dio 76,4; Herodian 3,12,9; A. Birley 1988, 162; Nero: above, pp. 78f; 18f. Young men: below, p. 28f.

77. Dio 76,14; A. Birley 1988, 183. It was not the first time Caracalla used soldiers of the guard for stirring up a disturbance, see Dio 76,3 and 76,14.

78. Caracalla's killings: Dio 77,1ff. Iulius Antonius: Speidel 1993, 69. Bears: Dio 77,10. Caracalla as a 'fellow soldier': Dio 77,13,1; drunk: Dio 77,17,4 and below, p. 138; cape: Herodian, 4,7,3; Dio 78,3,3; see below, p. 104. Unlike Augustus: Suetonius, *Aug.* 25; for emperors as 'fellow soldiers' see Campbell 1984, 32ff. 'His men': P. Giess. 40 (Oliver 1989), see Speidel 1978, 67 and 1992, 81.

79. Dio-Xiphilinus 78,4,1ff, the only authentic source according to Kolb 1972, 118f. Dio-Xiphilinus 78,5 describes the tribunes as belonging to the *doryphorikon*, which can mean the horse guard as well as the praetorians, see Dio 56,23,4.

80. Nemesianus: *AE* 1966, 606; Pflaum 1982, 316 a. Macrinus' faction had to hide after his death in AD 218, see Herodian 4,14,2 with Whittaker's commentary (1969).

81. Gothi gentiles: Speidel 1984, 254–8; Frisians, Tuihantes, *numerus Hnaudifridi*: *RIB* 1036; 1576; 1594; see Dio 77,14,3. Foreigners in the army: Speidel 1992, 71–81. Slaves as recruits: Suetonius, *Aug.* 25; in the higher

ranking branches: Digest 49, 16,2 (*specie militiae*). Labienus: above, p.14, also Petreius (Caesar, *BC* 1,75: *familia*); slaves in the horseguard, perhaps: Speidel 1993, 347. Compete: Speidel 1993, 415.

82. Ambassadors: Dio 78,6,1. Mauri of Lusius Quietus: *HA, Hadr.* 5,8, see Strobel 1984, 69; Mauri of Macrinus: Dio 78,11 and 32; compare also Tacitus, *Hist.* 1,59, and the Vandals of Stilicho. For Maximinus see below, p. 68; Caracalla's wig: Herodian 4,7,3. Willems 1884, 247 thinks Caracalla's guards spoke no Latin, but the translators Dio adduces were, more likely, for tribal ambassadors. Roman Germans in Caracalla's praetorian guard: Speidel 1992, 153ff; Batavians in his horse guard: Speidel 1993, 62. Carpi: below, p. 75.

83. *Germani corporis custodes*: Suetonius, *Cal.* 58; Septimius Severus had Narcissus thrown to the wild beasts because he had strangled Commodus (Dio 74,16). For the need to avenge see Tacitus, *Hist.* 1,40 (and 44): *scelus cuius ultor est quisquis*; see also Suetonius, *Claud.* 11; *Otho*, 1; *Vit.* 10; *Dom.* 23 and Dio 67,14,5. Dio 78,4 stresses Caracalla's death in the midst of the soldiers.

84. Battle near Antioch: Dio 78,37; Herodian 5,4,8. Verus: Dio 80,7; Speidel 1993, 69. Parade: Herodian 5,6,8. Avidius Cassius was killed by a centurion on horseback (*centurio exercitator?*), acting together with a decurion (Dio 71,27,3). The guards as masters of the emperors: Dio 79,17 = Petr. Patr. Exc. Vat. 152. Severus Alexander: *HA, Alex. Sev.*, 61–62; *Maxim.* 7; Herodian 6,8. For the discussion: Loriot 1974, 674f; Lippold 1991, 68ff; 383; compare Bang 1906, 87.

85. Citizenship in provincial towns was available for horseguards, see Speidel 1993, 15. Citizen of Nova Italica: Forni 1992, 206–10. Modern scholarship: Syme 1971, 179–93. Aquileia: Speidel 1992, 414ff. Herodian: 6,8,1–2 and 7,1. *HA, Maxim.* 2–4, see Lippold 1991, 80ff; 296; 304; 330f; Grosso 1966. *Turma Iuli Maximini*: Speidel 1993, 65 and 115. Syme 1971, 184 (grudgingly) and Whittaker 1970, 131 note 3, admit that the Augustan History may have information gleaned from Dexippus; Lippold 1991 is more willing to admit the use of other sources. Iulii in the Severan horse guard: Speidel 1993, 54; 56; 688–688e. The name may have been adopted upon joining the guard.

86. Horsemen of the *alae*, selection by the emperor: below, p. 78; *exercitator*-career: below, p. 110. Mauretania: *IAM* 298 (Sidi Kacem); for tribunes of the horse guard as governors of Mauretania see below, p. 100. Portrait: Fittschen 1980; Fittschen-Zanker 1985, no.105. Plate 6, photograph: K. Fittschen, Deutsches Archäologisches Institut, Athen. Ex-rankers rising to high office: below, p. 100f, see also Oclatinius Adventus, Dio 78,14,1, Rankov 1987. Forefront: Herodian 7,2,6–8. Medallion: Gnecchi 1912, 2, p. 87, no 4. Soldiers proud: Herodian 6,8,3; *HA, Maxim.* 7,1.

87. *Celeres*: Speidel 1992c, see *Damaszener Mitteilungen* 2, 1985, 12. Caracalla's Macedonians: Dio 77, 8f; Speidel 1993, 568. Maximinus: Herodian

6,8,3; 8,6,1; his birthplace: G. Forni 1992, 206–10. Senators as murderers: *HA, Maxim*. 18 see Lippold 1991, 520. Families: Herodian 7,3,6 and 8,5,8. For soldiers changing sides to protect their families see *Anonymus on Strategy*, 9 and 42 = Dennis 1984, 28f and 123f. Revenge: Herodian 7,11f.

88. Philip: Speidel 1993, 75. Altar: Speidel 1993, 54. Decius: A. Birley 1992 'Decius'. Onasander: 37,5. Parleys: Onasander 10,14 and below, p. 119f. Valerianus' capture: Kettenhofen 1982, 97ff; compare Tacitus, *Ann.* 13,38. Cameo: Göbl 1974.

89. Thessalonica: Speidel 1993, 75a; Zosimus 1, 43; Speidel 1992, 93. Good and evil done by soldiers: Speidel 1980, 39. Aureolus: Hoffmann 1969/70, 247; not commanding the whole battle cavalry: Simon 1980. Gallienus' horse: Hoffmann 1969/70, 243ff; Speidel 1984, 391–6. Faithful to Gallienus: above p. 102.

90. Altars in Rome: Speidel 1993, 70; 71. The horse guard as *comitatus*: Tacitus, *Ger.* 13, see above, p. 39f. Papyrus: P. Oxy I, 43, recto; the emperor in Egypt in 295 was Galerius, according to Barnes 1982, 62. *Comites*: Speidel 1992, 379–384 (add Lactantius, *De mort.* 38; Dexippus, *FGH* 100, F6, uses the term *taxis hetairike*, recalling Alexander's cavalry). Senator-*comites*: AE 1984, 145. Aristocratic standing: Speidel 1984, 245, with Hoffmann 1969, 79. *Equites promoti d.n.*: Hoffmann 1969/70, 246; Speidel 1992, 379–84. *Vexillationes*: Speidel 1984, 136–43. Caligula, competition: above, p. 23, also Vegetius 1,23 and 3,4; Tacitus, *Hist.* 3,27 and 29; Josephus, *BI* 3, 485 and 496; Aristides, *Or. Rom.* 85. *Doppeltruppe*: Speidel 1992a, 33; above, pp. 32 and 60.

91. For Galerius and Thessalonica see Laubscher 1975, 61ff with plates 45–50 and Barnes 1981, 18ff. Scorned armor: Vegetius 1,20. Mauri: see above, p. 41 and below, p. 84f; *cursores*: above, n.43. Marcus: Column, scene XVIII. Epona: below pp. 141f. City-Roman troops: Lactantius, *De mort.* 26,3. Praetorians with Diocletian in 303: Lactantius, *De mort.* 12. Praetorian tradition spurned: Mommsen 1910, VI, 234f; contra: Speidel 1992, 382. *Batavi* with Diocletian in summer 303: RIV 699. Rank and riches: Vegetius 2,24. Sex fiends, Carpi: Lactantius, *De mort.* 38 (*stipatores* are bodyguards, hence a horse guard; for *comites* see above, n.90, for *satellites* see Pliny, *Paneg.* 22). Carpi: Barnes 1982, 64 (they or their forerunners seem to be shown on Galerius's arch, frieze B,I,16, see Laubscher 1975, 45–8. Women: see Suetonius, *Aug.* 69 for what was customary to those in power.

92. *Comites*: above, pp. 73f. Batavi: Speidel 1993, 75a; *Notitia Dignitatum Oc.* VI, 47 and 51. *Scholae* as successors to the praetorians (*Saaltruppen* = standing ready for service in the palace lobby): Mommsen 1910, VI, 230ff; Grosse 1920, 58ff; 93ff; Hoffmann 1969/70, 279ff. Hoffmann's is the best account of the *scholae*. *Palatini* as successors of the praetorians: Mommsen 1910, VI, 234f. NCO clubs: Speidel 1993, 54; Rea 1984; almost units of their own: e.g. *CIL* VI, 32695 = Dessau 2791. *Schola*

armaturarum praetorianorum: *CIL* VI, 31122. *Scholae* as former NCO clubs: Frank 1969, 14ff (though p. 48 seen as newly founded by Constantine). *Scholae* not *numeri*: *Notitia Dignitatum Or.* 18,5; Mommsen, 1910, VI, 208, n.3. *Scholae* recruitment: Mommsen, 1910, VI, 232; Hoffmann 1969/70, 300f; Frank 1969, 59 and 67. Picked from regular units: *IK* 27, 95. *Protectores*: below, p. 131. Techniques, splendor: *schola scutariorum clibanariorum* (*Not. Dig. Or.* 11,8; SEG 20, 1964, 232), trained and paraded by Constantius (Julian, *Or* 1, 37–8; Ammianus 16,10,8) see Hoffman 1969, 265ff and 279ff.

Chapter 4 *Tall and Handsome Horsemen*

93. From tribal cavalry: Dio 55,24,7; perhaps *ala singularium* played a role, for it was under a Batavian commander (Tacitus, *Hist.* 2,22; 4,70; 5,21). Six feet tall: Vegetius 1,5. *Germani*: Speidel 1992, 112ff. Chosen from the *alae*: Speidel 1993, 63; 212; 541; 543; 603; 624; 641; 659; 663; 672. Cohortal horsemen second class: Dessau 2487 *equorum forma, armorum cultu pro stipendi modo*. Provincial *singulares*: Speidel 1993, 616, also ibid. 3, 22 and 624. Severus' legionaries: Dio 75,2. For *dona* see Maxfield 1981.

94. Years in the *alae*: Speidel 1965,2. Gravestones: Speidel 1993, p. 18; an exception is Speidel 1993, 266. For men dying at age 25 (due to age-rounding) see Scheidel 1992. Other specific lengths of service before transfer: Rea 1977, compare Tacitus, *Ann.* 15,59,4. Praetorians joining the guard: Kennedy 1978, 296.

95. Sons: Speidel 1993, 30; 169; 266; 599; 642; 660; 708; 709; 713; 733; 758; for the cavalry see also Speidel, 1992, 348, for sons among the praetorians Passerini 1939, 168. Fourth century: Frank 1969, 73ff. See also Cod. Theod. 6, 24,2; 7,1,8, etc. *Legio III Augusta* supplied itself with recruits see Le Bohec 1989, 521; for the army overall see Stein 1959, 59; Mócsy 1992, 38f. Sons training: Speidel 1993, 528, see below, p. 136.

96. Right to choose: Speidel 1992, 160. Caligula: Suetonius, *Cal.* 43, see above, p. 22. Candidus: Speidel 1993, 65. War games: Dio 59,2 (Caligula); Pliny, *Paneg.* 13; Marek 1993, 107; Fronto, *Princ.* 13. See below, p. 109f. Trajan: Pliny, *Paneg.* 15. Hadrian in Lambaesis: *CIL* VIII 2532 = 18042 with Le Bohec 1989a, 372; see above, pp. 47ff. Maximinus: *HA, Maxim.* 1,6. For third-century detachments of *alae* see Speidel 1984, 142f. Groups (*conalares*): Speidel 1993, 63; 648.

97. Strength: Tacitus, *Ann.* 15,66; Appian, *BC* 4,7. *Forma*: *HA, Maxim.* 3,6. For beauty in the selection see Onasander 10, 14; Herodian 4,7,3; 5,4,8; 6,8,1; Hoffmann 1969, 300 and II,126; Speidel 1992, 158. Juvenal: Satire 14, 194. Paphlagonian gravestone: Marek 1993, 107. Tallest: Herodian 5,4,8. 178. 6cm: Vegetius 1,5. European guards: Mansel 1984, 102f. Guardsmen tall and young: Vergil, *Aen.* 1, 497; Josephus, *BI* 5,460. Strongest: above, p. 18; Josephus, *Ant.* 19,1,15; Bellen 1981,42. For Germans see Caesar, *BG* 1,31; *B.Afr.* 40; Tacitus, *Ger.* 20 *haec corpora*

quae miramur. Caracalla's blond wig: Herodian 4,7,3; compare Arrian, *Tact.* 34,4. Nero and Vestinus: Tacitus, *Ann.* 15,69; cf. *Ger.* 13. Same age: Josephus, *BI* 5,400. Maximinus: *HA, Maxim.* 3,6. Marius, too, had chosen the fittest for his horse guard (Sallust, *Iug.* 98). Fear: Maurice 10,1; Onasander 37,5. For such criteria see Speidel 1992, 124ff. Zosimus 1,52.

98. Caesar: Suetonius, *Iul.* 65 and 72. Aristeides, *Or.Rom.* 74–8. Legionary recruits: Speidel 1984, 60f and Le Bohec 1989a, 530. Auxiliaries: Gilliam 1986, 202ff. Drifters: Tacitus, *Ann.* 4,4; see also Brunt 1974 and MacMullen 1990, 225f. Marcus: *HA, Ant. Phil.* 21. Fugitives: Digest 49,16,4. Dregs: Mócsy 1992, 35f. Keep them straight: 'Caesar', Dio 43,18,2 and 'Maecenas', Dio 52,27,4–5. Hardships: Aurelius Victor, *Caes.* 39,26. Mobs: *Bulletin Épigraphique* 1941, 92a, compare Caesar, *B.Afr.* 54.

99. Cheruscan prince, Parthian: above, p. 26f. Slaves: above, pp. 14; 66f. Fourth century: Frank 1969, 70f. Poem: *Anthologia Palatina* 7, 363, see Cumont 1942, 433. Written by Hadrian: see Soranus' poem, above, p. 46, *HA, Had.*, and Arrian, *Parth. frg.* 36 (Roos); very similar is Speidel 1993, 757; for the genre see Marek 1993, 107. Veteran horsemen as local worthies: *CIL* III, 770; 865; 1100; 15205,3; *CIL* V, 5006; *CIL* VIII 9062; 21039; 21064; compare Speidel 1992, 185; Forni 1992, 63; Passerini 1939, 166f; above, p. 93. Lurius: above, p. 53. Local aristocrats as common soldiers: Speidel 1993, 757; Passerini 1939, 164f; Forni 1953, 124f; Speidel 1984, 60f; SEG 1989, 1766. Aristocratic ideals: Devijver 1989, 279f. For officers see below, pp. 91f. Sons becoming knights: Speidel 1993, 34; 730.

100. Plate 7: Speidel 1993, 558, photograph B. Malter; parade: above, p. 50. Tacitus's overstatement: above, p. 28. Dio's overstatement: above, p. 58, see Durry 1938, 256f; Forni 1953, 123f. *Barbarica legio*: *CIL* V, 923, see Passerini 1939, 163 contra Durry 1938, 250. Marcus not understood: Dio 71,5,3. For the Latin of the guardsmen see below, p. 130. Nobility: see the *Celeres*-knights above, p. 70. Galerius: above, p. 74.

101. Tencteri: Tacitus, *Germ.* 32; see also Tacitus, *Ger.* 32 and Seneca, *Ep.* 36,7. Customs lingered: see Ammianus 25,6,14. Batavians best horsemen: Dio 55,24,7; Plutarch, *Otho,* 12; Speidel 1991. Young men in arms: Alföldy 1968, 94; Brunt 1975. For the data of fig. 5 and of the table see Speidel 1993, 14–17; coming all from the same cemetery they are unbiased and statistically valid: our sample compares well with 0.15% for the legions and 0.68% for the urban cohorts, see Forni 1992, 16. Statistics: Speidel 1965, 19; since 1965 the data has grown by ⅓. Pride: see the envoys sent from home, above, pp. 18 and 68; standard: compare *AE* 1948, 190 = *IAM*(aroc) 100. Aristides: *Or. Rom.* 74.

102. Shift: noted by Mommsen, *GS* 6, 1910, 71; Strobel 1989, 113. Severus dependent on the army: Dio 75,2. Honor of serving in the guard

bestowed: Cheesman 1914, 82. The Pannonian army pressuring the emperors: Mócsy 1992, 59–84. Constantine's *scholae*: Hoffmann 1969, 299f.: Strength of the *alae* in Germany and Pannonia (1500 and 7000 men): Cheesman 1914, 149 and 153f. Recruitment of the guard: Speidel 1993, 16. Conservative praetorian recruitment: Durry 1936, 256. Pannonians fine horsemen: Tacitus, *Ann.* 15,10; *Hist.* 3,2. Origin of praetorians: Passerini 1939, 171ff; of legion II Parthica: Forni 1992, 120; Speidel 1992, 39; transfers to legion II Parthica: *IGBulg.* III 2, 1570, see Dobson 1978, 313ff. British guardsmen: Speidel 1993, 107; 343; 652; see 742. Italy's horsemen: Junkelmann 1991, 41.

103. Afri: Speidel 1993, 208; 338; 639. Mauri: Speidel 1993, 34; 210; 607; 660; 755; below pp. 141; 73f; for their being spearmen see Speidel, 1992c. Parthian training: Herodian 6,5,4. Hadrian: Arrian, *Tact.* 44. Easterners: Speidel 1993, 44; 265; 510; 684. For Syrian archers in the Roman army see Kennedy 1989. Commodus: above, p. 53. Thracians, too, were tall: *maximi viri* (above, n.1). To Syrian soldiers those from Germany looked *truces corpore*, 'uncouth', perhaps 'tall' (Tacitus, *Hist.* 2,74). Herod: Josephus, *BI* 1, 673 and *Ant.* 15,9(12),3. Iuba: Caesar, *BC* 2,40,1.

104. A recruitment balance was advocated by Mommsen 1910, *GS* 5, 409; keeping the corps apart: ibid. GS 6, 70. For Noricans see Speidel 1982; Thracians in the fleet: Reddé 1986. Praetorians from all legions: Passerini 1939, 182f.

105. Provinces: Speidel 1993, 60 and pp. 14ff; Speidel 1992, 313ff. Ethnic groups: Dessau 2555; 9132; *RIB* 1538; 2148; Speidel 1992, 310. Thracians: Speidel 1993, 11; 26; 62 (see also 265). Hadrian: Arrian, *Tact.* 44, see below, p. 149f. For the growing specialization of fighting techniques see Speidel 1992a, 32. Frisians: Speidel 1993, 101; 180; 359; Macedonians: ibid. 568; 595; Batavians: 642; see also 169.

106. Domaszewski 1908, 196. Dio, consul of 229: Dio 80,5. Forerunners, axes, Commodus: above, p. 52. Galerius: above, p. 75. Fourth century: Frank 1969, 59f; Hoffmann 1969, 299f.

107. Greek name: *Gamus*, above, p. 25. Different names: see the Aelii of Speidel 1993, 13–15. Ambattus, Durze, Decibal: Speidel 1993, 236; 285; 60; 58. Soldiers not blending into the city: Dio 75,2,5; above, p. 60. Joining a legion: Forni 1953, 103ff. Citizenship for *auxilia*: Kraft 1951, 100ff; imperial names with citizenship: Dio 60,17; Kraft 1951, 101f; Speidel 1965, 61ff; Alföldy 1968, 105ff; no citizenship because of the horse guard diplomas: Mócsy 1992, 188ff; Eck-Wolff 1986, 567–575. Diploma with *et peditibus*: Speidel 1993, 79. Pisonian conspiracy, gravestone: above, p. 29; freedmen, Latin citizenship: Pancierra 1986.

108. Tradition: Gellius 10, 28. Hadrian: *HA*, *Hadr.* 10. Laws: see e.g. Codex Theodosius 7,22,2 and Codex Iustinianus 10,55,2; those over 40 years of age could no longer carry bow and arrows besides their regular

equipment: Maurice 3 1,2,52; a *senex* discharged after only 20 years of service: Rea 1984. 18 years minimum: P. Ryl. IV, 609 (a. 505) as quoted by Ravegnani 1988,21. The source for the table is Speidel 1993, p. 21. Age-rounding: see Scheidel 1992; to his survey of skewed recruitment-age statistics add Holder 1980, 123f; see also Forni 1992, 5, 18 and 67. Our percentages are calculated with the formula of Duncan-Jones 1977, 348f. If the second-century horsemen adjusted the years of service to fit a standard recruitment age, we could still say that the standard was 18–20.

109. Life expectancy in Rome: Clauss 1973b. Bellen 1981, 77–81, concluded, based on 11 gravestones, that the *Germani corporis custodes* died 'alarmingly' young. Since their situation differed little from that of the *equites singulares Augusti*, his sample seems too small; Kennedy 1978, 278 suggests that younger recruits died earlier, but this is not true for the horse guards where those recruited at age 16–17 served an average of 15 years, just like the rest. For the data see Speidel 1993, p. 20.

110. Forts: below, pp. 126ff; air: Hyginus 21; water: Colini 1944, 358; baths: Johnson 1987, 240ff; their own baths: below, p. 127; doctors and hospitals: Speidel 1993, 9; 31; 42; 268. Vitellius' army: Tacitus, *Hist.* 2,93–94. Plate 9: Speidel 1993, 531, photograph Musei Vaticani IV–30–7. 70 men needed: if the survival rate was about 54%. Decurions, centurions: below, pp. 147f. Survival estimates for praetorians: Kennedy 1978; cf. Angeli Bertinelli 1974.

111. Suetonius: *Aug.* 49; misunderstood: Raaflaub 1987, 247, but see Behrens 1986, 152. Veterans in the Batavian uprising: Tacitus, *Hist.* 4,70; robbers: Cod. Theod. 7,20,7 (a. 353). Gradations of reliability (in service): Hyginus, *De mun.* 2. 25 years of service: Speidel 1993, 731; 734; 736. *Ex causa*: Speidel 1993, 7; H. Grassl 1985; *CIL* XVI,10. Old age: Rea 1984. Down to days: Speidel 1993, 11. Poem: Speidel 1993, 596. *Labor*: Horsmann 1991, 187–97.

112. *Allegere*: see the inscriptions listed in n.93 and Suetonius *Aug.* 49; also Speidel 1993, 64. Officers: Alföldy 1987, 329. Diplomas: Speidel 1993, 76 and 79. Known promotions in the horse guard are as follows: *beneficiarius* to *sesquiplicarius* (Speidel 1993, 325; 433); *beneficiarius* to centurion (688c; 732); *duplicarius* to decurion (679; 700; 744, also 691; 692); *tablifer* to decurion (40); *signifer* to *trierarchus* (753); to decurion in the provinces (15; 37; 60); decurion to *decurio princeps* (22; 23); for decurion to legionary centurion see below, p. 147f. Careers of *singulares*: Domaszewski 1908, 50–3; Speidel 1965, 22ff; Breeze-Dobson 1993. Imperial appointment: below, p. 96. Under-officers: Speidel 1993, 56; Arrian, *Tact.* 42; Speidel 1992, 62ff; room: Hyginus 7; followers: Vegetius 2,19; punish: Digest 49,16,9. Standards followed: Arrian, *Tact.* 35,2. *Beneficiarii*, other ranks, and promotion criteria: below, p. 96. The centurionate as the soldiers' dream: Alföldy 1987, 31; *Curatores*: Speidel 1993, 54; 56; 58. *Vexillarius*: Speidel 1993, 57; 113;

114; 326; 710; see Domaszewski 1885,25; Speidel 1965, 40f; Marichal, *ChLA* 8, p. 19. Hope: Aristides, *Or. Rom.* 85. Altar: Speidel 1993, 63 (the *tablifer* suggests they are close to retirement). Towards the end: there are few under-officers' gravestones (Speidel 1993, 11).

113. For promotions from decurion to *centurio exercitator*, and from there to the Rome tribunates, see Speidel 1993, 30; below, p. 111. Praetorian soldiers could become tribunes by the mid-first century: Dobson 1978, nos 69; 70; 78; A. Stein 1927, 160ff; for horse guardsmen see Speidel 1992, 296f and Devijver 1992, 331ff. A guardsman like Maximinus, could *impetrare ab imperatore quod vellet* (*HA, Max.* 3,6).

114. British: Farwell 1989, 179–90 (the Tamils of Ceylon have now turned from fainthearted to ferocious). Clerks: P.Berl. inv. 11596R = *BGU* 1689 = *ChLA* X 422, see Speidel 1992, 320. Nations: Vegetius 1,2 *gens gentem praecedit in bello.* Egyptians: Strabo 17,1,53; Syrians, Jews, Cilicians: Dio 71,25. Asians etc: Seneca *De ira* 1,11,3f; Greeks: Tacitus, *Hist.* 3,47. Italians etc: Dio 75,2,4. Germans: Tacitus, *Ann.* 54 and *Ger.* 13–14; Josephus, *Ant.* 19,1,15 (122, see above, p. 24); Seneca, *De ira* 1,11,3f. Pannonians: Herodian 2,9,11. Thracians: *Expositio totius mundi* 50 (=Riese, *Geographi Latini Minores* 117) *maximi viri et fortes in bello*; Ammianus 26,7,5.

115. Second-century horsemen: Speidel 1993, 12–17. For Raetians see below, p. 113. Pannonian and Norican under-officers: Speidel 1993, 315; 340; 530; 545a; 752; 165; 114; 337. Thracian under-officers: Speidel 1993, 63; 108; 258; 296. Praetorium: Durry 1938, 248; fleet: Reddé 1986, 532. Fine Thracian horsemen: Kiechle 1964, 117–22. Seneca: *De ira* 1,11,3f. For the ethnic role of under-officers see also Domaszewski 1908, 72; Vegetius 1,2. *Fortia facta*: Speidel 1992, 124–8. Principle: Aristides, *Or. Rom.* 85; followed: Speidel 1984, 56 (Aurelius Atticus).

116. Discharge dedications: Speidel 1993, 1–19. Hadrian knowing the names: *HA, Hadr.* 20 (it was a *topos*, see Pliny, *Paneg.* 15,5). Silvanus: Speidel 1993, 688d, compare Alföldy 1987, 450 whence *AE* 1978, 440. Retentus: Speidel 1993, 68, see also Domaszewski 1908, 41 and 54, and Le Bohec 1989, 204 and 211; add Dessau 505. Compare *evocati*, E. Birley 1988, 326–330 (*iteratus*: *CIL* VI 2534 = Dessau 2050); veterans to teach experience: Maurice, *Strat.* 2,7. See also CIL V, 8744 and 8755.

117. Veteran settlement: E. Birley 1988, 272–283. Bellen 1981, 79 feels the *Germani* regularly returned home. Galba, Severus: above, pp. 29f; 57f. Law: below, p. 132. Veterans' gravestones from Rome: Speidel 1993, 728–38 (add 659 and 701); Italy: 739–41 (add the poem above, p. 46); Britain: 742(?); Thrace 743; Asia Minor: 757. Diplomas: 76–9, 758; 759. Third-century praetorians: Roxan 1981; legion II Parthica: Mann 1983, 48. Praetorians in country towns: Durry 1938, 303. *Veteranus Augusti*: Speidel 1993, 25; 735; 736. Aristocrats: Speidel 1993, 730. Veterans in high society: Forni 1953, 42ff and 1992, 33f and 79f; *AE* 1915,69; see also Speidel 1993, 757 and above, n.99. Zenodotus, above, p. 80; see

also Speidel 1993, 757 and above, n.99. Furlough: Speidel 1992, 341; contra: Forni 1992, 75. Cash awards: Suetonius, *Gal.* 12; Speidel 1992, 368; Caesar, *B.Afr.* 40. No *praemia* for auxiliaries: advocated by Wolff in Eck-Wolff 1986, 48ff.

Chapter 5 *Aristocratic Officers*

118. For the rank of the tribunes and their command see above, pp. 59f. Aemilius Macer: Digest 49,16,12; see Devijver 1989, 1–15. Pliny, *Paneg.* 18. Discipline in the city: Val. Max II,7 introd. and II,7,5. *Emansores*: Digest 49,16,3. An altar to Disciplina from Birrens/Scotland (*RIB* 2092) shows the fort's walls and closed gate. Scourge: Tacitus, *Ann.* 1,20 and 1,36; P. Dura 55; Speidel 1984, 308f. Endless work: Tacitus, *Ann.* 1,35, also 1,31; Maurice VIII,2,15; many duties: Forni 1992, 27. Death penalty: Digest 49,16,3. Dress: Vegetius 2,12. Jealousy: Pliny, *Pan* 18.3 Wife: Tacitus, *Ann.* 2,55 and 3,33; Dio 59,18,5.

119. *Decurio princeps* and staff: Speidel 1993, 23;54; Speidel 1965, 35f; 1984, 189–95. *Adiutor praetorii*: Speidel 1993, 634. For the ranks mentioned see Speidel 1993, index; *medicus*: 31 and 42; *optio valetudinarii*: 9 and 261; *tablifer*: 7; 8; 9; 16 (veterans); 40; 63; Huelsen *CIL* VI, 31152 suggested the *tablifer* was a standard bearer, he was followed by Domaszewski 1908, 51, Durry 1938, 31; Speidel 1965, 37f. Emperors appointing under-officers: Speidel 1984, 180; 1992, 126. For *beneficiarii* see Speidel 1993, index; for their tasks Domaszewski 1908, 21; dedication of a *schola*: Speidel 1993, 54. Caesar: *B.Afr.* 54. Perks: Tacitus, *Hist.* 4,48. Brave deeds: Speidel 1992, 124–30. Killing panthers in the circus: Suetonius, *Claud.* 21.

120. Aurelian panels: Ryberg 1967. Plate 4: photo Archivio della Soprintendenza Archeologica di Roma. Officer mistaken: Ryberg 1967, 28ff, see Speidel 1993, 419. Officers holding flags: Tacitus, Ann. 1,38; Suetonius, *Aug.* 10. On such occasions the emperors rode white horses: Pliny, *Paneg.* 22. *Hastiliarii*: below p. 130. For the emperor's battle standard see pp. 27 and 120; the emperor giving the sign with the flag: Caesar, *BC* 3,89; *B. Alex.* 45.

121. For this uniform see Speidel 1993, *Anhang*. Helmet: Devijver, 1989, 390ff. Dear: MacMullen 1960,24 on P. Princeton 57; gift: *HA, Hadr.* 1,17,2. The coat is the same as that of the equestrian officers on Trajan's council of war, Trajan's Column, scene VI. Sword not very sharp: Tacitus, *Ann.* 1,35. Another tribune of the *equites singulares Augusti* is shown, very likely, on the Borghese battle sarcophagus (Schäfer 1979) wearing almost the same dress with *caliga* boots and, it seems, the unit's leaf-pattern shield, for which see below, p. 105. Killing officers: Tacitus, *Hist.* 3,31. Emperor's dress: Dio 75,1; see Plate 20. Dress slashed: Frontinus, *Strat.* 4,1,28, see E. Birley 1988, 155.

122. For *adventus* see above, p. 41f. Coins: Trajan: *BMC* 511; 1014 (a. 114/5); M. Aurelius: *BMC* 1349; 1373; 1030; 1068–71. Commodus: *BMC* 1707;

see Halfmann 1986, 143–148. Medallion: *BMC, Medals, Marcus Aurelius, AE* 12. *Parma equestris:* Devijver 1992, 350.

123. Pliny's praise: *Paneg.* 18 and 19. Sweat, blood, etc: Pliny, *Ep.* 2,7,1. Tribunes of the *Germani:* the title *chiliarchos* (above, p. 24) often means tribune. Gladiators: above, pp. 25 and 27; Domitian: above, p. 33f. For equestrian officers see Devijver 1989 and 1992; for *primipilares* Dobson 1978; for praetorian officers Durry 1938, 127ff.

124. Three years as the usual length of service for commanders of frontier cavalry regiments: E. Birley 1988, 150. Independent administration: Alföldy 1987 3, 109. One step below praetorian tribunes: Speidel 1993, 72; 73; 74. For the role of the praetorian prefect see e.g. Reddé 1986, 517f; also Durry 1938, 165ff; Passerini 1939, 266ff; Speidel 1965, 28f. Dio 52, 24 suggests all troops stationed in Italy be subject to the praetorian prefect, but this is a suggestion for Caracalla, not a statement of fact, see Millar 1964, 115. Praetorian prefects are mentioned Speidel 1993, 11; 14; 15; 19; 54; 57; 59; 60; they are lacking on Speidel 1993, 16; 30; 69. Perhaps the prefects also chose horse guards to serve as *protectores* on their staff: Speidel 1993, 543; 578; 580.

125. Tribunes: Speidel 1993, 72; 73; 74; 11; 14–16; 30. Marcius Turbo, and tribune of 120: Speidel 1993, 72 and 74. *Magister equitum*: Hoffmann 1969, 301. Cleander: Herodian 1,12,3; see Grosso 1969, 227ff; above, p. 52.

126. Third-century tribunes: Speidel 1993, 55; 54; 55; 57; 60; 69; 71; 75a; 76; 78; 79; 115; Dio 78,5 (Nemesianus and Apollinaris). Domaszewski (1908, 137; 164; 189) took a certain Iustus and his colleague of *CIL* VI, 1110 to be tribunes of the Mauri and Osrhoeni; more likely, they are tribunes of the horse guards, here mentioned because they were the superiors of those *frumentarii* who served as the horse guard's trainers (Speidel 1993, 34; 753; 754; 755). Equal rank: above, p. 59f. For Aurelius Nemesianus in Mauretania see above, p. 65.

127. For troopers becoming officers see Speidel 1993, p. 3, and above, p. 69 and 91f. Riding skills needed by tribunes: Dio 77,8. The boyhood training of equestrian officers is well described and illustrated by Gabelmann 1989; see also Forni 1992, 69–71; Brunt 1975, 270. Cassius Dio: 52,25,6–7; 52,26,1. The *Troia* was not military, see Kiechle 1964. Middle Ages: Hale 1983, 255ff. Third-century officers from the ranks: Dio 78,13f; A. Stein 1927, 164ff; Syme 1971, 190; Loriot 1975, 668; *AE* 1980, 705; *Britannia* 20, 1989, 333; E. Birley 1988, 34; Speidel 1992, 296; Devijver 1992, 316–38; above, p. 69. Corruption: MacMullen 1988, with my remarks in the *American Historical Review* 1990; old ethics: E. Birley 1988, 153.

128. Recruited less from the elite: Alföldy 1987, 35ff; Devijver 1992, 316–38. Trajan: Pliny, *Paneg.* 19. Battle sarcophagus Borghese; see Schäfer 1979, 361: 'Kommandant einer Reitertruppe'. Though a prefect of an

ala cannot be excluded, the scale mail and the proximity to the emperor suggest the horse guard. The three Romans wearing coats clasped in the middle of the chest and muscle cuirasses, one of them offering a helmet, may be fellow officers as on the relief of Mikkalos, Devijver 1989, 390ff.

129. For the house see Colini 1944, 350–2. At the legionary fortress of Lauriacum in Noricum the houses of the tribunes lay even farther away. Living standards in the field: Caesar, *B.Afr.* 54; Velleius 2,114. Soldiers as servants: *Ad obsequia*: Vegetius 2,19. Officer's pay: Devijver 1989, 409. British in India: Farwell 1989, 93. Trooper's pay: below, p. 133.

130. *Curam agens*: Speidel 1993, 22. *Curam agentes* and *praepositi* in the *auxilia*: E. Birley 1988, 221–31. Petronius Taurus: Speidel 1993, 75. Valentinus: Speidel 1993, 75a; Dio (Boissevain) III, pr.743,162, see above, p. 72. *Vir perfectissimus*: Speidel 1993, 71; 75a; compare Speidel 1993, 75 and *AE* 1975, 882; whole house: above, p. 59. Tribune in 286: Speidel 1993, 71. Constantine's commander: *AE* 1975, 882, likewise governor of Mauretania Caesariensis; Pflaum 1975 argues he headed Mauretania first and the guard afterwards, which is quite possible, considering the guard's growing importance; the title is up from *sacri lateris custos* (Martial 6, 76).

Chapter 6 *Weapons and Warfare*

131. Trajan's Roman-ness: Pliny, *Paneg.* 13,5 (though even his army dress was mostly German to believe Tacitus, *Hist.* 2,20). For the soldiers' dress see Ubl 1969; Bishop-Coulston 1993; Speidel 1993, p. 8f. *Sagulum*: Tacitus, *Hist.* 2,20. Linen: Colton-Geiger 1989, 55; goat leather and knees covered: Maurice, *Strat.* 1,2. Plate 10: Speidel 1993, 529, photograph Musei Vaticani IV,30,1. *Paenula*: Speidel 1993, 90; 114; 509; against rain: Dio 57,13,5; *HA*, *Hadr.* 3. Second-century *sagulum*: Speidel 1993, 109. Caracalla: Dio 78,3; there is no need to rely on Herodian's more explicit but less reliable report (4,7,3). *Sagum* German: Tacitus, *Ger.* 17. Northerners needed: Tacitus, *Ann.* 3,40.

132. Diocletian's edict: Lauffer 1971, *sagum* and *equus optimus militaris*. Caracalla: Herodian 4,7,3 (if to be trusted). Tribune: Vegetius 2,12. Shining arms and armor: Onasander 10,14 and 28; Vegetius 2,14; Josephus, *BI* 5, 350–6; Maurice, *Strat.* 1,2,25. According to Tacitus, *Ann.* 1,24, untidy looks betrayed insubordination (*contumacia*), which was a punishable crime. Free dress and food: Vegetius 2,19. Enlarged guard: Rostovtzeff 1957, 510. Galerius: Lactantius, *De mort.* 37 and above, p. 75.

133. Spears used for thrusting and throwing: Arrian, *Tact.* 4,7 and Tacitus, *Ger.* 6. Plate 8: Speidel 1993, 83, photographs Musei Vaticani XXXV–4–11–3 and 4. *Lanceae*: Balty 1987; Dessau 9134. Speidel 1992a, 14ff; 1993, 686. *Pilum*: Speidel 1993, 580; ibid. p. 9.

134. *Spatha*-sword (70cm; 3 pounds): Connolly 1987, 17; slashing sidewards: Arrian, *Tact.* 43 used in case of need: Tacitus, *Hist.* 1,79; Arrian, *Tact.* 4,3. Flank attacks: Arrian, *Ect.* 31; *Tact.* 4,9; finishing off enemies: stele of T. Claudius Maximus, Speidel 1984, 178 and the stele of unknown origin Schleiermacher 1984, 83. *Beaux sabreurs*: see Speidel 1993c; perhaps Elagabal was one such: Dio 78,38,4. British enemies: below, n.16; Germans: Tacitus, *Ger.* 6 (though cavalry gravestones on the Rhine show enemies with swords); Mauri: Speidel 1984, 140–1.

135. Medium cavalry: as defined by Asclepiodotus 1,3. Bows and arrows: Speidel 1993, 14; 510; 684 (plate 3); laths: Bishop-Coulston 1993, 135ff. Axes: Arrian, *Tact.* 4,3 and *Ect.* 31 (for the *alae*, hence also for the guard); *Not. Dig. Oc.* 9,2; Turks in modern times still used such Amazon weapons; add axes – Leo, *Tact.* 6,33 mistook them for maces – and the *gaesum*-spear to Bishop-Coulston 1993, 165 etc.). Crossbows: Arrian, *Tact.* 43; Coulston 1985, 259–63; compare Chinese mounted crossbow-men, Ames 1993, 29; D.B. Campbell 1986. *Architecti*: Speidel 1993, 9; 223; 321; they are weapons engineers, see Onasander 42,3 and Ammianus 24,4,8. Slings: Arrian, *Tact.* 43, see below, p. 113. Rocks: above, p. 18. *Gaesum*: below, p. 113. Mamluks: Ayalon 1961. *Tectores*: Speidel 1993, 64; see the *tegentes*, Ammianus 19,7,8; also cavalry *defensores*: Maurice, Strat. 2,3; Anonymus Byz. 17 (Dennis 1985, 57f), and the *catafractarii* in the *alae*, Speidel 1992, 409.

136. Shield-size (1.1m): Groenman-van Wateringe 1974. Fourth-century shield badges: Ammianus 16,12,6; Vegetius 2,18; Speidel 1992, 414–18 (Tacitus, *Hist.* 3,23 refers perhaps to color). Round shield: Speidel 1993, 686 (a Norican), see Speidel 1993c; for *lanciarii*-horsemen in the guard see also *CIL* VI, 32965 = Dessau 2791, Hoffmann 1969, 218f. A *lanciarius* with a round shield: Balty 1987, fig. 5; *gaesati*: below, p. 113. Choice armor: *Paneg. Lat.* 12,5,3, see below, p. 154. Emperors wearing scale: Schumacher 1982,44; Fittschen-Zanker 1985,92. Heavy armor hated: Fronto, *Princ.* 12; Vegetius 1,20. Bare hands: Fronto, *Ep.* II,1, van den Hout 122.

137. Fig. 7 was drawn by P. Connolly, London, after Plates 9, 15, and 20. Eagle-head swords: plate 14 and Speidel 1993, 565. A helmet rent by a spear: Caesar, *B.Afr.* 78; see Arrian, *Tact.* 40. Gala helmets: Speidel 1993, 544; 568; 596; 598; 682. Silk, etc: Claudianus, *Sixth Consulship*, 575ff. Parade weapons a gift from the emperor: Suetonius, *Caes.* 67, see Herodian 2,13,9; Dio 78,6; Nuber 1972; Speidel 1992, 131–6. Worn on parades only: Josephus, *BI* 351. Sports armor: Arrian, *Tact.* 34, see below, p. 113f.

138. Saddle and harness: Speidel 1993, 9f; fine illustrations: Connolly 1988, 31. Celtic-type saddlery: Bishop 1988. Quality control: Speidel 1984, 330. Felt blanket: Frontinus, *Strat.* 2,4,6; and Diocletian's price edict s.v. *centuclum*. Feathers plucked: Fronto, Ep. II,1 (van den Hout 122).

139. The relation of pay and weapons: *CIL* VIII 18042 = Dessau 2487.

Deduction for the repair of arms: P. Vindob. L. 72 (= Fink 1971, no.71): *refec(tio) arm(orum)*. For the horse guard's weapon keepers on lists and gravestones see Speidel 1993, 12; also Speidel 1992, 131–6. One in each *turma*: Speidel 1993, 56. The last datable weapon keeper of the horse guard is Speidel 1993, 60 of AD 205; in the army at large the rank is still found under Gallienus: *AE* 1936,55. Ranking of branches: Digest 49,16,2; Speidel 1992, 307. Ownership and issue of weapons: Speidel 1992, 131–9; Bishop-Coulston 1993, 198–201.

140. Losses: P. Dura 97 = Fink 1971, 83; Tacitus, *Ann.* 2,5. Horses shot first: Caesar, *BG* 4,12. Doubling: Walker 1973, 341. 30% replaced: Hyland 1990, 86; Junkelmann 1991, 112. For cavalry horses see Walker 1973; Davies 1989, 153ff; Junkelmann 1989; Hyland 1990. Breeding grounds (Apulia, Campania, Reate): Hyland 1990, 12f. Gravestones: Speidel 1993, 675 (Stabiae, Campania); 676 (Lucera, Apulia); 680–1 (Rubi, Apulia); 677 (Reate); perhaps also 678 (Capena, north of Rome) and 679 (Interamna, Umbria). Dio: 52,30,7. Stallions: Plate 3; kick and bite: Hyland 1990, 80f; no stallions: Davies 1989, 168. Bringing down foes: Speidel, 1993, 540; compare Josephus, *BI* 3,488; *Paneg.Lat.* 4(10),29,5; or stelae like that of Romanius Capito, *CIL* XIII 7029 = Schleiermacher 1984, 27 = Boppert 1992,135.

141. Prices: Gilliam 1950, 180f (with little regard, however, for inflation) and Hanson 1985. AD 139: P. Yale 249 = Fink 1971, no.75; AD 208: P. Dura 56 = Fink 1971, 99; AD 251: P.Dura 97 = Fink 1971, no 83 = ChLA 7, 325. For the pay of the *auxilia* see Speidel 1984, 88 and M.A. Speidel 1992, with a confirmation of the 5/6 scale by a new pay record from Vindonissa. The price for horses going up with the pay: *CIL* VIII 18042 = Dessau 2487. Diocletian's price edict (Aizanoi fragment) values an *equus optimus militaris* at 3600 *denarii*; a soldier's pay then was 700 *denarii stipendium* plus 6250 *donativa* (Jahn 1984, 58). Praetorian: Dio 75,1. *Curatores*: Speidel 1993, 54; 56; 58; 139; Speidel 1992, 137. Gravestones: Speidel 1993.

142. Napoleon: Chandler 1966, 351. Highly trained: Caesar, *BG* 1,39. For the strategists' 'projection of force without its actual use' see Dio 52,37 and 69,9; above, p. 47; Josephus, *BI* 3, 108 and 475f; Suetonius, *Cal.* 19,3 with Dio 59,17,7 on the Puteoli bridge, see above, p. 21; Aurelius Victor, *Epitome* 14,10. Strategic force: contra Campbell 1984, 114. Finest horsemen: Dio 55,24; for the skills the guardsmen brought along see above, p. 78. See also Velleius 2,109 on Maroboduus' horse guard: *perpetuis exercitiis*. Fewer = better: Frontinus 4,3; *Paneg. Lat.* 12,5; Aristides, *Or. Rom.* 86 and 77; Vegetius 1,1 and 3,1. Rome ruled because of training: Vegetius 1,1; Josephus, *BI* 3, 102ff; on the training ground: above, pp. 27; 78. Domitian, Trajan: Pliny, *Paneg.* 13.

143. Pliny: *Paneg.* 13; see above, p. 42. For *centuriones exercitatores* see Speidel 1993, 11; 14; 15; 19; 30; 34; 54; 55; 57; 60; 69; 753–6; also Maximinus, above, p. 68f. Domaszewski 1908, 106 thought they were praetorians,

yet many, perhaps all were *frumentarii* (Speidel 1993, 34; 753–5). Roman-ness alone seemed poor to Arrian, *Tact.* 32ff. Drillmasters in other units: Speidel 1978, 28ff and above, pp. 34; 49; they were of lesser rank, see Horsmann 1991, 81–107; Durry 1938, 116f. Shaping tactics: see below, p. 149f. Training officers from decurions: Speidel 1993, 30, compare Speidel 1965, 43. Emperors tried: Arrian, *Tact.* 32ff, see Kiechle 1964, also Dio 69,9. Severan training officers: Speidel 1993, 54; 55; 57; 60; 69. Caracalla bent on murder: above, p. 64f.

144. Bowmen: Speidel, 1993, 14 (?); 510; 684; see also *CIL* VI, 3595 (*vitis*, hence *evocatus* or centurion); Speidel 1965, 45; Durry 1938, 117; compare Vegetius 1,15 and Barsemis Abbei at Intercisa, Speidel 1984, 146; Coulston 1985, 289. *Ala milliaria*: Hyginus 16, see Lenoir 1979, 66ff. Guard decurions set over 30 men: Speidel 1993, 58; compare the many fourth-century centurions Vegetius 2,7 and *RIU* 2, 559. *Turmae* in training and in the field: Arrian, *Tact.* 35 and 42. Duties of the decurion: Vegetius 2,14 (he is wrong saying a *turma* had 32 men, see Speidel, 1992a, 28). Plate 12: Speidel 1993, 30, photograph Musei Vaticani IV, 30, 8. *Virga* originally for auxiliaries: Livy, *Per.* LVII. Thrashing: Tacitus, *Ann.* 1,23; Apuleius, *Met.* 9, 39–40; Marichal 1992, 82. Punishments: Dig. 49,16,13. Horses sweating in battle: Festus M. 221; spares in battle: Maurice 3,8. Decurions rose: Speidel 1993, 30, and 733; above, p. 91f. Fell in battle: Speidel 1993, 756. Maximinus: above, p. 69.

145. For training see Hyland 1990 and Horsmann, 1991. Vaulting into the saddle: Fronto, *Ep.* 2,1 (Van den Hout 122); horses running: Arrian, *Tact.* 43,4; older men: Vegetius 1,18, see 2,14. Tribunes: Dio 77,8, see above, n.127; Emperors' grooms (*stratores*): *HA, Car.* 7.

146. Spear throwing: Horsmann 1991, 149–54. Remain seated: Arrian, *Tact.* 38,3. Little Brits: Vindolanda tablet T 1985/32 (*AE* 1987, 746; R. Birley 1990, 26): – *Brittones. Nimium multi equites. Gladis non utuntur equites nec residunt Brittunculi ut iaculos mittant.* Pliny, *Nat. Hist.* 8,65. Spears: Fronto, *Ep.* 2,1 (Van den Hout 122): *Haud multi vibrantis hastas, pars maior sine vi et vigore tamquam lanceas iacere* (javelins = *lanceae*: Balty 1987; Dessau 2487 and 9133–5; Speidel 1992a, 14ff). Index finger: Trajan's Column, scene LXIV. Foot soldiers stabbing: Schleiermacher 1984, 29; Boppert 1992, 62. Attacks: Arrian, *Tact.* 36ff, see Lammert 1931, 46–59; Kiechle 1964; Junkelmann 1991, 175ff. Wheeling left, unknown to Germans (Tacitus, *Ger.* 6, see Lund 1991, 2050f), was also rare among *equites cohortis* (*frequens dextrator*, Dessau 2487): Lammert 1931, 55. Flank attacks: Arrian, *Ect.* 31.

147. Rules, Arrian, etc, see below, p. 149. Arrian, *Tact.* 43–44. Crossbow: above, n.135. Throwing rocks: above, p. 18 and below, p. 138. Horsemen as slingers: *CIL* VIII 18042 = Dessau 2487; Watson 1969, 60f; Griffiths 1989; Völling 1990. Griffiths, 269 suggests horsemen alighted for slinging. Arrian's 'Celts' and 'Getae' are Germans and Dacians when he speaks of his own time, see his *Ect.* 2 (contrast 9) and

8, also Arrian, *Parth.* 36 (Roos); *Anat.* 1,3,2, he calls the Quadi 'Celts'; contra: Kiechle 1964, 114. Quadi, *contarii*: Speidel 1992, 64–6. *Gaesum*: Helbig 1908, 5f. Quadi, Dacians, Raetians: see *ala I Ulpia contariorum, ala I Ulpia Dacorum*; *cohors I Aelia Gaesatorum milliaria* in the new Pannonian diploma of 125; Speidel 1981. Raetians not promoted: Speidel 1993, 84; 85; 88; 109; 133; 140; 297; 234; 287; 324; 329; 672; 683; 553; 581; 640; 644; 654. Sports armor: Garbsch 1979; no more than a face mask has been found in Rome (Garbsch 1979, O–11). Cavalry shows: H.A. *Hadr.* 9,8; below, p. 132.

148. For the *campus* of a unit see Horsmann, 1991, 60ff. Civilian: Cod. Iust. Inst. 4,3,4. Britain, Syria: Horsmann 1991, 75ff. Tor Pignattara: Colini 1944, 419; Guyon 1987; Speidel 1993, 30.3km as in Tarraco (Speidel 1978a, 72). *In Comitatu*: Depositio Martyrum (AD 354, going back, perhaps, to 311: Mommsen Chron. Min. I,72 = *MGH AA*, 9); Seeliger 1987; for the use of *in comitatu* meaning 'with the guard' see *AE* 1949, 38; Tacitus, *Ger.* 13; *Hist.* 1,23; the horse guard as *comites*: above, p. 73ff. Sceptical about a second field of Mars as a training ground: Deichmann-Tschira 1957, 66f and Guyon 1987, 422. Daily exercise: Vegetius 2,9. Cemetery of the *Germani*: Bellen 1981, 56 and 62. Onasander: 10,6; Vegetius 3,2 (see also 1, 27) and Horsmann 1991, 176. Tiber: Speidel 1991; Horsmann 1991,130.

149. Pliny: *Paneg.* 13. Emperors training: above, p. 27 (Nero); 53; see also Suetonius, *Aug.* 83; *Tib.* 13; Plutarch, *Pomp.* 41,4–5. Public watched: Suetonius, *Ner.* 10. Domitian: Suetonius, *Dom.* 19 and Pliny, *Paneg.* 13. Reputation needed: Pliny, *Paneg.* 17; Vegetius 3,26. Maxentius: *Paneg. Lat.* 12,14,4. See also LeBohec 1989, *Armée* 114f; the topic would repay more research. Swimming the Tiber: below, p. 155.

150. Hyginus 7;8;30, see Speidel 1993, *Anhang*; Lenoir 1979, 58. Gerasa: Speidel 1993, 22. Altar for Hercules: Speidel 1993, 55. Whole unit: see also Speidel 1992, 380f. Older men: Speidel 1993, 688,d, compare van Rengen-Balty 1992, 30.; left to guard forts: Tacitus, *Hist.* 4,14. Gravestones abroad: Speidel, 1993, 682–90; 688–688e (a guardsman's gravestone does not prove the emperor was there, though). Tiberius: above, p. 19f. Greet the emperor: above, p. 97f.

151. Aristides: *Or. Rom.* 80. 'Long road': Tibullus 1,26, see also Speidel 1980, 39f and 1992, 201f. Travel 65km: Hyland 1990, 163; 95km. Gelzer 1960, 273. Unshod: Hyland 1990, 123f. Caesar: *B.Alex.* 9 and 34. Trajan's Column: scene xxxiv. Domitian's and Trajan's river trips: Pliny, *Panegyric* 81f; Strobel 1984, 104 and 244; 1989, 83; Arrian, *Parth.* frag. 62 and 65 (Roos). For a seagoing horse transport, a converted trireme, see Hyland 1990,98; Caesar used freighters to ferry the horse to Africa (*B. Afr.* 2) and the ships even served as stables (*B.Afr.* 7). Sea and Neptune: Speidel 1993, 25. For Septimius Severus sailing see *HA, Sev.* 15,2. For warships in Asia Minor, Noll 1990, 257. Sea travel preferred: Halfmann 1986,65ff; sea-sick: Caesar, *B.Afr.* 18: *ex nausea*.

152. Grooms: below, p. 136f. In camp: Hyginus, 7. Grooms on campaigns: Speidel 1993, 688b; 688e; 690; Speidel 1992, 65 and 342–52; Ammianus 19,8,7; decurion: Speidel 1993, 688b, cf. O. Bu Njem 86. Forage: Maurice, *Strat.* 7,B,10. Law: Cod. Iust. 12,35,12. Lovely boys: Speidel 1993, 691. Helmets: Speidel 1993, 544; 568; 596; 598; 682; spear: 595.

153. To the risks the emperors ran, described by Campbell 1984, 65ff, add scouting; compare Caesar, *BG* 1,22. Suetonius, *Cal.*45 see also above, p. 19 and 22f. Scouting by commanders: Tacitus, *Hist.* 5,1; Festus, *Brev.* 25; Maurice, *Strat.* 8,2,65; Nikephoros, *Skirmishing* 12 (= Dennis 1985, 187); enemies' weapons: Arrian, *Parth. et Suc. Al.* frag. 21 (Roos). Caesar in his tent: *B.Afr.* 31. Energetic commander: Campbell 1984, 35. Reasons why a commander could not trust reports are given by Tacitus, *Ann.* 2,12. Scouting by day: Aeneas Tacticus 6; at night: Vegetius 3,6. Crowns abolished: they are lacking in Pliny's account, *Nat. Hist.* 16,3 and 22,4; Maxfield 1981, sadly, has nothing on these crowns.

154. *Cura explorandi*: Tacitus, *Hist.* 3,56 and Devijver 1976, V,23, described by Maurice, *Strat.* 9,5. *Speculatores*: above, p. 33ff. Titus: Josephus, *BI* 5,52 and 5, 258. Trajan's Column: scene 58. Alexander: Arrian, *Alex.* 3. Best horses for scouts: Vegetius 3,6; Maurice, *Strat.* 9,5. For *exploratores* generally see Speidel 1992, 84–104. Law: Digest 49,16,5. Onasander 10,14. For handsome guards see above, p. 78f; their weapons: above, p. 104. Caesar, *BG* 1,42. Valerianus: above p. 71.

155. Otho: Tacitus, *Hist.* 2,33, see above, p. 34. Banner as point of reference: Herodian 5,49, see also Dio 40,18; Ammianus 16,12,39; Maurice, *Strat.* 1,2,80; Durry 1938, 203; contra: Zwikker 1937,12. Sign for the attack: Caesar *BC* 3,89; *B.Hisp.* 28; *B.Alex.* 45. Borne by the guard: see the bearers of Constantine's *labarum* (Frank 1969, 142ff; Pflaum 1975) and above, pp. 27 and 98. Commander's place: Vegetius 3,18; Maurice, *Strat.* 2,16.

156. General not to fight: Onasander 33, yet if need be: Tacitus, *Hist.* 2,5; Trajan's courage was famous (Dio 68,6). Pliny: *Paneg.* 17. Compare Tacitus, *Ann.* 2,20 and Suetonius, *Cal.* 3 on Germanicus. Septimius Severus: above, p. 61. For emperors in battle see also Campbell 1984, 59ff; *Paneg. Lat.* 4(10)29,5; 12(9),3ff; above, pp. 69; 72; and below, p. 154; fourth century: Frank 1969, 100f. Duties: Onasander 33; see also Arrian, *Ect.* 22f; Maurice, *Strat.* 8,2,100.

157. Napoleonic cavalry commanders: Chandler 1966, 355f. Thapsus: Caesar, *B.Afr.* 83. Reining in: Maurice, *Strat.* 12,B,17. Titus: Josephus, *BI* 3,483. Elagabal: above, p. 67. Horse bolting: Tacitus, *Agr.* 37,6.

158. Self-sufficient units: Vegetius 2,2; also Maurice, *Strat.* 2,8. Hadrian: Dio 69,9. Reserves: Tacitus, *Ann.* 2,45; Ammianus 25,5,9. Forward attacks: Tacitus, *Ann.* 2,65 (Arminius *cum delectis*). Shoring up the line: Dio 75,6,6; Vegetius 3,18 (*supernumerarii*); Speidel 1978, 48ff. Maurice, *Strat.* 8,2,100; Tacitus, *Hist.* 3,23. Marius, Jugurtha: Sallust, *Jug.* 98.

Caesar, Titus, Severus: see above, pp. 12; 35, 61; Arrian *Ect.* 22f. Self-sufficient units: Vegetius 2,2; also Maurice, *Strat.* 2,8. Killing those who fled: Frontinus, *Strat.* 2,8,14; Appian, *Lib.* 124. Flight: Zonaras 12,24 and below, p. 154; cf. Tacitus, *Ann.* 14, 32; likewise the guards of the Numidian, Parthian, and Iberian rulers: Sallust, *Jug.* 54; Tacitus, *Ann.* 6,35. 'Peak of success': Tacitus, *Hist.* 2,24.

159. Swimming: Tacitus, *Hist.* 4,12, see also Bang 1906, 39 Nile: Caesar, *B.Alex* 29; Danube: Speidel 1991 and above, p. 46. Tiber: below, p. 154. Romans dreaded swimming, Germans tall: Tacitus, *Hist.* 2,18. Uncontested crossings: Vegetius 3,7. Seine: Piekalkiewicz 19 80, 31. Scatter: Caesar, *B.Alex.* 29. Learn to swim: Horsmann 1991, 127–32, Tacitus, *Ann.* 14,29; *Agr.* 18. Trajan's Column: scene 31. Arminius and Flavus: Tacitus, *Ann.* 2,9 (above, p. 26); bowmen of the guard crossing first: above, p. 46. Titus: Josephus, *BI* 3, 497. Maximinus: Herodian 7,2,6ff; above, p. 69.

160. Spanish horsemen: Dessau, 8888, see Lammert 1931, 3. Caesar: *B.Alex.* 17. Iotapata: Josephus, *BI* 3, 254 – the men may have been Vespasian's horse guard, see Speidel 1994. Caesar, *B.Afr.* 29; Herod in 37 BC attacked Jerusalem with his *singulares* (*epilektoi*): Josephus, *BI* 1,37, and P.Ostorius in Britain in 50 did a similar thing (Tacitus, *Ann.* 12,31). Mural crown: Suetonius, *Aug.* 25; Aulus Gellius 5,6,16; Maxfield 1981, 64 and 77 feels taking a town was not needed to get such a crown. Hatra: Dio 68, 31, see above, p. 45; see also Ammianus 19,7,8. Plate 11: Speidel 1993, 90, photograph DAIR Inst. Neg. 63. 3068. A few hundred: Dio 75,12,5. Attributing deeds to the leader: Tacitus, *Ger.* 14; for Rome compare Tacitus, *Agr.* 8,3; Josephus, *BI* 3,298; 324. Batavian *virtus*: Tacitus, *Ger.* 29. Street fighters (bowmen): Josephus, *BI* 5,331–41; Suetonius, *Tit.* 5. Septimius Severus: Dio 75,11,3, see above, p. 62. Riding up to a wall: Caesar, *B.Afr.* 29. For Spanish horse guardsmen see Caesar, *BC* 1, 75; Harmand 1967, 459ff, though neither Harmand nor Fröhlich 1882 mention Strabo's Spaniards. Honor and pay spurred: compare Josephus, *BI* 5, 464.

161. Spilling non-Roman blood: Tacitus, *Ann.* 14,23; *Agr.* 35, see Walser 1951, 39. Frontier troops' opinion: Tacitus, *Ann.* 1,17. Battle decorations of praetorian and legionary horsemen: Maxfield 1981, 216f; Schleiermacher 1984, 96; Speidel 1994. Papyrus: P. Ross. Georg. III,1, see Speidel 1993, 415f. Fight: see the *exercitator* killed in battle above, p. 111. For doctors assigned to the horse guard see above, p. 88f. Auxiliary units' battle awards: Maxfield 1981, 218ff; individual auxiliaries: Maxfield 1981, 52 and 54. Life in Rome: Herodian 1,6,7; Suetonius, *Titus* 5; Lactantius, *De mort.* 26: *ibi vivere optarent.* Thanked: Speidel 1993, 58.

Chapter 7 *Life in Rome*

162. Old Fort: Colini 1944, 314–17. Two dedications: Speidel 1993, 20; 21; see also 1. 'Noble hall': below, p. 139. Figure 10 follows Colini 1944, Tav. XXIIII. Further west: if so, the long marble-floored walkway west of the road belongs to a later time; further north: Speidel 1993, 45. Engineers: Hyginus 56. Sejanus: Tacitus, Ann. 4,2. For a wall to keep soldiers in, see above, p. 95f. *Campus Viminalis*: Durry 1938, 54f; Tacitus, *Ann.* 12, 36. Statues: Colini 1944, 315ff.

163. New Fort: Colini 1944, 353–9; above, p. 59. Plate 13, photograph Pontificia Commissione di Archeologia Sacra, Lat. G. 27. *Curatores*: Speidel 1993, 54 and 58. Strength: above, p. 59. Troop of 30: above, p. 111. Cellars: Johnson 1987, 191; veranda: Colini 1944, fig. 294, right; *HA, Hadr.* 10. Like homes: Tacitus, *Hist.* 2,80; Livy 44,39,5. Fort wall, cistern: Krautheimer 1977, 29. Gravestone relief: Speidel 1933, 595; praetorian merlons: Johnson 1987, plate 2,c. Diplomas: Speidel 1993, 76–9; 758. Forts for units: above, p. 73. Baths: Pellicioni in Pietri 1976, 6.

164. Bodyguards: *Germani* and *singulares* shared the name *Batavi*: Speidel 1993, 688–688e; above, p. 62; Herodian, too, calls the *singulares* consistently 'bodyguards': 1,8,6; 3,12,9; 4,13,1; 5,4,8 (in battle!); 4,7,3; 6,8,7, see also Dio 61,9,1; Grosso 1964, 34; above p. 52. Augustus: Suetonius, *Aug.* 19; Seneca, *Clem.* 1,9,6. Assassins: Tacitus, *Hist.* 2,75; Dio 74,16. Oath: Epictetus, 1,14,15. Nero: Suetonius, *Ner.* 47,3, see Bellen 1981, 92. A building: Bellen 1981, 44 and 57 (*Wachlokal*); hallway: Isidore, *Etym.* 9,3,42: *in porticibus excubant*; rooms for praetorians in the palace: Durry 1938, 56f. Night duty: above, p. 54. Drowsy: Herodian 2,1,2; below, p. 138. Death penalty: Digest 49,16,10. Frontier soldiers: Tacitus, *Hist.* 1,17.

165. Groomed: above, pp. 81; 128. Hair knot: Tacitus, *Ger.* 38; Much 1967, 428. Praetorians beards: Galba's *adlocutio*-coin Ryberg 1967, pl. xxxviii. No Latin: Hirschfeld 1913, 588; Willems 1984, 247 (as for Dio 79,6, the ambassadors, not the guardsmen, needed translators). Language of the inscriptions: Speidel 1993, pp. 17 and 25; see also Tacitus, *Ann.* 2,10; 2,13,2 (a horseman); Le Bohec 1989a, 546f. For Tacitus calling them 'foreigners', see Walser 1951, 67–72. Poetry: Speidel 1993, 596; 692; 760; 761; above, pp. 46; 136f. Caracalla: below, p. 138.

166. Crowds: Herodian 2,6,13; Tacitus, *Ann.* 12,43. Eyes: Herodian 3,2 (Plautianus). Low-ranking officers: Tacitus, *Hist.* 2,88. Drillmasters: Speidel 1965, 44f; Horsmann 1991, 84–92. Domaszewski 1895, 93, took *hastiliarii* for elite horsemen. Training spear: Vegetius 2,15; 4,8; 4,18; Onasander 10,4; Arrian, *Tact.* 34,8; 40,4; for this meaning of *hastile* see Speidel 1965, 44f; 1978, 30f; 1984, 330f; add *Sammelbuch* 11641; cf. Apuleius, *Met.* X,1. Davies 1989, 88 saw that crowd control gear resembled practice weapons; see also Speidel 1993b. Use of spear-butts: Livy 35,5,10. *Speculatores*: above, pp. 33ff. *Hastiliarii* inscriptions: Speidel 1993 7 (AD 135); 8; 9; 11; 12; 13; 15; 16; 17; 110; 287; 325; 332;

342 (too many for being drillmasters); see also above, pp. 43ff. A governor's *hastiliarius*: CIL VIII, 2562 = 18051. Napoleon: Chandler 1966, 354 (the largest field army under Trajan was at most 200,000 men: Strobel 1984, 154); above, p. 34. *Protectores*: Speidel 1993, 543; 565; 580; ibid. p. 9; Speidel 1986; *protectores* in the provincial guards: Speidel 1978, 130–3; idem 1994. *Pilum*: Speidel 1993, 580; ibid. p.9.

167. Other escorts: Tacitus, *Ann.* 13,18,4; Suetonius, *Nero* 34; Dio 61,8,6. fourth century: Frank 1969, 100; Millar 1977, 65f. Hunting: Devijver 1992, 140–7; Speidel 1993, p.7. Gravestones outside Rome: Speidel 1933, 671 (Ostia) 672 (Albano); 673 (Albano); 674 (Tivoli); 675 (Stabiae); a *speculator* at Anzio: *AE* 1955, 24 (above, n.36).

168. Military pomp: Suetonius, *Nero* 13. A ceremony with parade helmets is shown in the Aurelian 'prisoners' panel, Ryberg 1967, plate xxxix and p.58. *Adventus*: above, p. 42. Pay-day parades: Josephus, *BI* 5, 348ff. State funerals: above, pp. 50; 58f.

169. Claudius: Suetonius, *Claud.* 21; on African beasts see Devijver 1992, 141; Nero: Dio 61,9,1. Titus had 5000 beasts killed in one show, Suetonius: *Titus* 7; Trajan 11,000: Dio 68,15. For soldiers and amphitheaters, see LeRoux 1990. Durry 1938, 58f suggests that the *vivarium* at the Porta Labicana belonged to the horse guard; for soldiers keeping game, see Devijver 1992, 140–7. Mock battles: Speidel 1978a, 43; cavalry battles in the circus: Suetonius, *Dom.* 4 and Dio 67,8; see also above, p.114.

170. Bowmen: Speidel 1993, 510 and 684. M. Antony's archers: Cicero, *Philippics* 2,19; 2,112; 5,18; 13,18. Handiness of bowmen in street fighting: Josephus, *BI* 5, 331–41; for their use generally see Coulston 1985. Governors: L. Robert 1963, 363. For Herodian's (1,12) false report of a battle between the people and the horsemen see Speidel 1993, 412f. Spearmen against civilians at Carthage in 238: Herodian 7,9,7. Commodus' return: Herodian 1,7,6.

171. Jealousy: above, p. 66. Civilians against praetorians: Juvenal, *Sat.* 16, see Durry 1935; battled: Dio 73,13,4f; 80,2,3; Herodian 7,7,1ff; 7,7,5ff; Aurelius Victor 40,24. Heirs: Speidel 1993, 530; 543; 547; 578; 606; 608; 736; 744–50; 758; Speidel 1993, p.22f; Durry 1938, 248; Passerini 1939, 190f. Praetorianus: Speidel 1993, p. 36; Batao: van Rengen-Balty 1992, pl.3. Bulla Felix: above, p.63, compare Dessau 2488.

172. Allurements (*inlecebra*): Tacitus, *Ann.* 4,2; Vegetius 1,3. Serve in Rome: Suet. *Titus* 5; Herodian 1,6,7; see the motto to this chapter. Dishonorably discharged: Digest 49,16,13; Dio 75,1; Herodian 2,13,9. Deserters: *Dig.* 49,16,5,3. Lazy etc.: Tacitus, *Hist.* 2,21; also Herodian 2,10,6; *HA, Did. Iul.* 5,9; *Paneg. Lat.* 12, 21,2,f; 14,6. Theater: Tacitus, *Ann.* 13,24–25. Cures: Tacitus, *Ann.* 13,35; Fronto, *Princ. Hist.* 12.

173. Pounce: Tacitus, *Ann.* 1,20 and 1,36; P. Dura 55; Speidel 1984, 308f; above, p. 52f. Death penalty: Digest 49,16,16. Go free: Juvenal, *Sat.* 16.

Provinces: Pliny, *Ep.* 10,78. Axes: Herodian 2,4,1 and 6,10. Punishments: Digest 49,16,3. *Nota:* Tacitus, *Hist.* 1,52. Stricter: Digest 49,16,10. Marriage: above, p. 134f. On crime and punishment generally 'see Campbell 1984, 300-14.

174. Wisdom: Vegetius 3,26. For pay figures see M.A. Speidel 1992; there is no support in the sources for Domaszewski's notion (1908, 51) that *duplicarii* of the horse guard became decurions of the *alae*, see below, p. 147. If a horse guard decurion got four times basic pay (16,800), his pay still rose upon promotion to legionary centurion (18,000). Loot: Tacitus, *Hist.* 3,17; Dio 74,8; Vegetius 2,24. Cash bonuses: above, p. 30. Caracalla: Dio 77,24. German guards' expectations: Tacitus, *Ger.* 14. Caligula: Josephus, *Ant.* 19,1,15 (121). Gravestones: above, pp. 21; 25; 144; 1000 *denarii*: van Rengen-Balty 1992, 34. Miserly pay: Tacitus, *Ann.* 1, 35; squandered: below, p. 156.

175. Heirs: Speidel 1993, 21ff. Speidel 1992, 132. The data from the gravestones of the *auxilia* are slim here and allow no comparison. Shady dealings with the money a soldier hoped to get back for his horse: Hanson 1985. Pliny: *Ep.* 7,31. *Germani* casting dread: above, pp. 24f. Soldiers recognized: Herodian 7,11,2. Boors: Dio 75,2, see above, pp. 58; 81; 157.

176. Africa: Lassère, 1977, 481; Le Bohec 1989a, 542f; Mogontiacum, Carnuntum: Alföldy 1987, 33f; *Auxilia*: Roxan 1991; horse guard, second century: Speidel 1993, 313(?); 318; 722; third century, ibid. 528; 530; 535; 539; 544; 567; 570; 579; 594; 622; 624; 634; 635; 649; 651; 652; 654; 658; 724-6. Families lessening the soldiers' usefulness: Herodian 3,8,5; hampered mobility: Tacitus, *Hist.* 2, 80; Herodian 6,7,3.

177. Such a case is found with the forerunners of the horse guard, the former auxiliaries among the Flavian praetorians, for a diploma of 71 grants marriage rights to those among them who had married a non-Roman woman. CIL XVI, 25: [– *qui uxores habent aut qui uxores non] habent, siqui eorum feminam peregrinam duxeri(n)t* in Nesselhauf's reading. See Passerini 1939, 40f; Lieb 1986, 324; above, p. 32. For *Batavi* married before and hence during service in Rome see Roxan 1985, 149; Bellen 1981, 63f; 79f; 103.

178. Soldiers originally forbidden to marry: CIL XVI, p. 154 (Nesselhauf); Mirkovic 1980. Praetorians forbidden to marry: Durry, 1938, 289ff; Lieb 1986, 344f; Behrens 1986, 165; buying houses: Digest 49,16,13. *Severior disciplina*: Vegetius 2,3. Ranking the branches of service: Speidel 1992, 307. For a third-century wife brought from Thrace, see Speidel 1993, 544; compare 761; for a praetorian's wife brought from Germany, Speidel 1992, 153ff. Families went along when auxiliaries were transferred, see Roxan 1978, *RMD* 86 (but as a privilege: Cod. Theod. 12,35,10). Wierschowski 1984, 298 suggests few horsemen of the *alae* had wives because they, too, were elite soldiers, but the evidence is too meager; see also Roxan 1990.

Gravestones by slaves etc: Speidel 1993, p.23. The wives (Speidel 1993, 722–6) do not show the same family names as their husbands and thus are not the horsemen's freedwomen.

179. Soldiers of legion II Parthica: Herodian 8,5,8 with Whittaker's (1970) comments. Wives younger – general population: *ZPE* 42, 1981, 60; *auxilia*: Roxan 1991. Horsemen's gravestones set up by wives: Speidel 1993, 318; 528; 530; 535; 539; 567; 570; 579; 624; 634; 635; 649; 654; 658. Gravestones for wives: Speidel 1993, 723 and 726 (plaques); 722 and 725 (stones with reliefs); 724; 761. 'Sweetest': Speidel 1993, 635; 'well deserving': 634. Poem: *AE* 1966, 22 (Auluzanus could be the horseman of Speidel 1993, 54, or the freedman of 313). Funerary banquets with wives: Speidel 1993, p.5.

180. Gravestones set up by children: Speidel 1993, 594; 622; 651; 652. Daughters of auxiliaries: Roxan 1990. Further relatives of the guardsmen: Speidel 1993 p.21, and 708–27. Astonishingly, uncles and cousins are all from the mother's side, if we may trust the terms *avunculus* (Speidel 1993, 718; 719, cf. 599) and *consobrinus* (596; 655), (see Garnsey-Saller 1987, 145) – does this confirm Tacitus, *Ger.* 20,3? Plate 14: Speidel 1993, 528; photograph Musei Vaticani IV–30–3. For soldiers shown with sons see Noelke 1986. Throwing rocks: above, pp. 18 and 112f. Sons joining the guard: above, p. 78. Parents: Speidel 1993, 596; above p. 80, see Le Bohec 1989a, 169. Knight: Speidel 1993, 730.

181. Price of a boy: *AE* 1896, 21 = *FIRA* III, 132. Freedmen: Speidel 1993, 313; 701; p. 23. Wash the horses: above, p. 118; Longrein: Ammianus 19,8,7. Mischief: above, p. 96. Bought a wife: Speidel 1993, 701. *Vigiles*: Speidel 1993, 734. Same tribes: Speidel 1992, 360 and 1993, p.23. Aurelius Sanctinus' two gravestones: Speidel 1993, 691 and 692. Plate 16: Speidel 1993, 535; photograph École Française de Rome, Neg. PM 594. For the disc on the baldric see Bishop-Coulston 1993, fig. 12 and pp.130f. Joined in death: Speidel 1993, p.22.

182. Heirs: Speidel 1993, 19. Messmates: Eck-Wolff 1986, 5; black mark: Digest 49,16,5,6, compare Tacitus, *Ann.* 14, 27; mixed: Maurice 2,7. Wounding a fellow soldier: Dig. 49,16,6. Sex: Suetonius, *Dom.*, 10; see also Dio 67,11. Brothels: *Notitia regionum* (Richter 1901, 371 ff.); pubs: CIL VI, 9992. Dice: Caligula, Dio 59,22; Suetonius, *Dom.* 21. Fronto, *Ep.* 2,1 (van den Hout 122): dice and drinking everywhere. Vegetius 2,14 finds a decurion should be *sobrius*. Praetorians: Tacitus, *Hist.* 1,26; 1,80. Commodus: Herodian 2,1,2. Caracalla: Dio 77,17,4 (Nicomedia); above p. 65. German guard's banquets: Tacitus, *Ger.* 14. Nero: Suetonius, *Nero* 21. Royal guards: Mansel 1984, 143ff.

Chapter 8 *Gods and Graves*

183. Excavations: Colini 1944, 314ff. Plate 17: Speidel 1993, 16; photograph Soprintendenza Archeologica di Roma. Roman army religion: E.

Birley, 1988, 397–432. Youngest: Speidel 1993 63. Reliefs: Speidel 1993, 19; 43–5. Statues: Colini 1944, 317. Bacchus honored by Septimius Severus: A. Birley 1988, index. Meduris: Speidel 1993, 36, Beellefarus 38, Dolichenus 42, add Sabadius 63. All gods honored: Speidel 1978b. Cicero: *De harusp.* 19.

184. Veterans' monuments: Speidel 1993, 1; 2; 3; 4; 6; 7; 8; 9; 10; 12; 13 (AD 118–41). Individual dedications: 23; 25; 26; 29; Cocceius Firmus: *RIB* 2174–2178, see E. Birley 1961, 87–103 and above, p. 50. *Ceterisque Dis*: Speidel 1993, 9; 10; 12; 13; 26; 29. Domaszewski's (1895) interpretations of these gods are mistaken. Home gods of units: Ankersdorfer 1973, 215f; home gods in Rome: CIL VI, 2797–2860. German gods: Speidel 1993, p. 30, n. 182. Felicitas: Speidel 1965, 68ff. Groupings: Bauchhenss 1976; Latte 1960, 334; 337. Praetorians: CIL VI 2821 (=Dessau 2096); 2822; Dessau 4633–4635; Durry 1938, 332. Menmanhia: Speidel 1993, 49.

185. Eponae: above, p. 48. Plate 19: photograph Archaeological Museum, Thessalonike. The relief is 1.4 m long, see Laubscher 1975, 149. For Epona: Magnen 1953; Speidel 1965, 68ff; Speidel 1978, 39f and 60; add now *RIU* III, 869, *IMS* I, 171, and Kazarow 1938, no. 399 = fig. 224 (Augustae/Moesia); see Speidel 1993, 3; 4; 6; 7; 8; 9; 10; 12; 13; 23; 29. For Galerius, see the well-researched though bitter account by Barnes 1981, 18ff.

186. Matres: Schmidt, 1987, 146f. Campestres: E. Birley 1988, 433–5; Speidel 1992, 290–7. Brought to the provinces, Germany: E. Birley 1988, 435, no.31; Dacia: above, p. 48. Britain: E. Birley 1988, 434, no.4; Africa: Speidel, 1992, 290–7; Spain (Mars Campester): E. Birley 1988, 404. Feminine: Speidel 1993, 2; 33. Campestres and Epona together: Nigrinus' inscription, above, p. 48; also CIL III, 11909 = Dessau 4830. Speidel 1993, 33 shows Campestres associated with the Gallo-German Suleviae. Three Mothers: *RIB* 1334; goddesses of the horse guards: Speidel 1965, 55–7.

187. Matres Suleviae: Speidel 1993, 3; 4; 6; 9; 10; 12; 13; 23; 25; 29; 61. *Matres* lacking: Speidel 1993, 33;46, but see the comments on 33. Personal guardians: Speidel 1993, 61, see CIL XIII 5027. 'Well-guiding ones': Schmidt 1987, 149; other etymologies: Rüger 1987, 3. Good son: Speidel 1993, 61, see Latte 1960, 337. Caesar, *BG* 6,21 (Nemetes, ibid. 1,51,2); tradition: Herodotus 1,131; enlarged list: Speidel 1993, 25.

188. Shunned: *HA, Hadr.* 21. Under Marcus: Speidel 1993, 42; see also 43; 44 and 18; compare *HA, Marcus* 13. Following the emperor's lead: contra Frank 1969, 222. Mithraea: Colini 1939, 313; 356; 363. Plate 18: Colini 1939, 363; Helbig 1963, I, 746f; Vatican, ex-Lateran inv. no. 343; photo DAIR Andersen 26.376. Roman values: Speidel 1980; Clauss 1992, 277f (relation to soldiers). For *velocitas* as a virtue of the soldier see also Rebuffat 1984, 233. Christian praetorians: Mercurelli 1939 contra Durry 1938, 348ff. Horse guard unchristian: Ferrua 1949, 655ff. Gravestones

without DM: Speidel 1993, 525; 526; 535; 536; 570; 601; 624; 641; 658. Brothers (*in pace*): CIL XI 6168 = Dessau 9075 = Speidel 1992, 381, perhaps they came with Galerius to Italy in 307, see below, p. 154.

189. Genius on altars: Speidel 1993, 31; see Ankersdorfer 1973, 46–57. Thanking the Genius: Speidel 1992, 306–8. *Salvo domino*: Speidel 1993, 66. Dedications by the horse guard after Severus: Speidel 1993, 62–4. Antoniniani: Speidel 1993, 62 of AD 219; missing ibid. 63 and 64. For honorific titles see Fitz 1983; Speidel 1992, 306–8.

190. For gravestone scenes, see Speidel 1993, p.3–8, for beliefs ibid. and the wonderful pages of Nilsson 1974, 486–98ff. Spheres: above, p. 25. Second-century horse guard gravestones scenes come from the Rhine rather than from the Danube, contra Bishop-Coulston 1993, 26. Plate 15: Speidel 1993, 682, photograph Soprintendenza Archeologica Milano, Neg. D/1539; see below, p. 153. *Aper fatalis*: Mócsy 1977, 376 with fig. 2; *HA, Carinus*, 14–15.

Chapter 9 *Training Faithful Frontier Armies*

191. Problem: A. Birley 1976, 260. Double the pay: Dio 53,11,5; most faithful: Hyginus 2. Mention of *alae* in Rome: above, p.77. Frontier armies: Speidel 1993, 63; 659; 672; 722. Heirs: Speidel 1993, 109; 256; 641; see also 746. Dio 89,4f. For the power of the Pannonians in Rome, see also Mócsy 1992, 59-84. Goading: *Paneg. Lat.* 12,14,6 and 21,2,f; theaters etc: ibid. 21,3. Despised legions: Durry 1938, 250 – but the inscription CIL V, 923 on which this is based belongs to the second century when the praetorians came from Italy, not from the legions, see Passerini 1939, 163. No help: Campbell 1984, 386. Democratic trend: Speidel 1992, 124–8; 163ff; Lactantius, *De mort.* 19. Decorations: e.g. CIL XII, 2230, see Maxfield 1981, 110-144; Syme 1991, 106; punishments: Tertullian, *De cor.* 1,2. Delegations: Campbell 1984, 268f; transfers to the guard: Speidel 1992, 160; above, p.57f.

192. Need to bind after 69: E. Birley 1988, 203. Promoted to decurion upon discharge: Speidel 1993, 15; 60; perhaps also: 37; 688; 722. Title: Speidel 1993, 15. Appointment by governors: Gilliam 1986, 191-205; by emperors: Speidel 1984, 180; Campbell 1984, 105 sees this as exceptional, but more likely it was typical, compare Dessau 9173; *AE* 1985, 849 (read *provectus*, instead of *praefectus* and Marek 1991, 100ff. *Ala Siliana*: Tacitus, *Hist.* 1,70 and 2,17. Provincial guardsmen appointed: Speidel 1978, 68f. Discharge list: Speidel 1993, 60.

193. Centurions appointed: E. Birley 1988, 207; Speidel 1992, 126. Campbell 1984, 101–9 undervalues the emperor's role, see e.g. *HA Hadr.* 10,6, Martialis' centurionate (above, p. 66, Dessau 9173, Marec 1991, and Passerini 1950, 594. *Beneficium*: Sallust, *Jug.* 96; Speidel 1992, 126; 139; 308 above, p. 58. Records: Gilliam 1986, 191–205. Tiberius' centurions: Suetonius, *Tib.* 12. Many guardsmen promoted: Speidel 1993, 27; 29;

33; 48; 53; 53a; 68; 132; 732; 751; 752; 757; ibid. 14; 377; 414; also Speidel 1965, 47f. Promotions quick: above, p. 91. Chief centurions: Speidel 1993, 28; 754.

194. Britain: Tacitus, *Hist.* 3,44, see A. Birley 1990,11. For gaining loyalty through the appointment of centurions, see Tacitus, *Ann.* 4,2; E. Birley, 1988, 207; Speidel 1992, 124–8. For under-officers as opinion leaders, see also Elagabal's promise, Dio 78,32. Commodus: Speidel 1993a.

195. Uniform training needed: Vegetius 2,2. Aristides, *Or. Rom.* 84; Horsmann 1991, 172f. *Evocati*: Domaszewski 1908, 77; Durry 1938, 117ff; E. Birley, 1988, 326f; Horsmann 1991, 90. Important function: Durry 1938, 125f; Passerini 1939, 127; Speidel 1965, 58ff; E. Birley 1988, 203; contra: Dobson 1967, VIIIf and 1978, 59 (*non sequitur*); Campbell 1984, 114. Bring innovations: Domaszewski 1895, 33; 1908, 77. Recent works emphasizing the role of training: Le Bohec 1989, *Armée* 111–25, and the outstanding book by Horsmann, 1991. Praetorian weapons development: Baatz 1980; doubted: Passerini 1939, 74; weapons development in the provincial guards: Speidel 1992, 136.

196. Arrian: *Tact.* 32–44, see Kiechle 1964 and above, p. 112. *Constitutiones*: Vegetius 1,8; 1,27. *Disciplina*: Horsmann 102–7. For army commanders: Arrian, *Tact.* 42.

197. For Hadrian's personal interest in cavalry maneuvers see Dessau 2487 and 9133–5; Dio 69,9. War academy: Speidel 1965, 55–60. Campestres: Speidel 1965, 55–58; E. Birley 1988, 433–5; Horsmann 1991, 105f; Speidel 1992, 290–7. Horsmann 1991, 91, against Speidel 1965, 57f, is right in that the *auxilia* had various kinds of training officers, see also Speidel 1992, 146 and 168, and Speidel 1993, 641. For commanders of *auxilia* see E. Birley 1988, 221–31; esp. Cocceius Firmus, 229 (see above, p. 50f; for an erstwhile horseman as prefect of an *ala*, see Speidel 1992, 296. Thanks to Campestres: Speidel 1993, 27; 33; 37. Trainers of provincial horse guards: Aurelius Decimus, CIL II, 4038 = Dessau 2415 = *IR*Tarraco 38, see E. Birley 1988, 4, 404 and 421; also Calventius Viator, above, p. 47ff. Provincial guards as training schools: Speidel 1978, 51f. Augendus: Speidel 1992, 296; Iulianus: Speidel 1984, 341–7 (he took the recruits from Rome to Caesarea as he travelled to his new job); idem 1993, 356. Maximinus: above, p. 68f.

198. *Evocati* from *auxilia*: CIL XIII, 1041; AE 1976, 495. Hadrian's *auctor*: HA, *Hadr.* 10, see Leglay 1974, 278. Training since Hadrian: Domaszewski 1895, 33 and Durry 1938, 124. Disciplina worshipped since Hadrian: Horsmann 1991, 102ff. 'Glad to see slip': Pliny, *Paneg.* 18,3. Mamluk: Ayalon 1961, 57. Chinese horse: Han-dynasty bamboo strips from Ta t'ing county (Ames 1993, 258). Fourth-century cavalry one-third: Hoffmann 1969–70, 193; *scholae* guards: above, p.75. Hyginus' *Camp Layout*, written, perhaps before Trajan's Parthian war, still assumes whole units, but see e.g. CIL VI, 32933 = Dessau 2723 = Saxer

1967, 44. Pius' Mauretanian War: Speidel 1984, 211–14; 1992, 64–6. Mt Amanus: Dio 74,7, see Speidel 1984, 219 and E. Birley 1988, 24. 'Secret': compare Frank 1969, 2.

Chapter 10 *Death at the Milvian Bridge*

199. Gravestone (motto): Speidel 1993, 756. *Remansores*: Speidel 1992, 280 and 1993, 57; dedication to Minerva: Speidel 1993, 54. Diocletian: Aurelius Victor 39,46f. Praetorian cohorts: Forni 1992, 415f. Galerius: Lactantius, *De Mort.* 26, see above, p. 73ff. For the uprising see Speidel 1992, 279–89. *Turmae praetoriae*: Aurelius Victor 40,5; the phrase may hark back to Tacitus, *Ann.* 12,56 which, in turn, is a pluralization of Suetonius, *Claud.* 21, for the praetorians had neither the *turmae* nor the maniples Tacitus mentions (contra: Durry 1938, 99); for *turma* in the fourth century meaning 'cavalry unit' see Speidel 1992, 282f; any unit called *turma*: Codex Theodosianus VII,13,8 (a. 380). *Subsidia*: Caesar, *BG* 1,83; Tacitus, *Ann.* 4,73; *Hist.* 2, 24; Vegetius 2,2 and 3,17; Aurelius Victor 40,25, see Speidel 1992, 282.

200. Squandering: *Paneg. Lat.* 12(9),3. Other emperors: above p. 104. Constantine's *liberalitas*: ibid. 15; Maxentius' *latrocinium*: ibid. 12(9),17. Gravestone: Speidel 1993, 682.

201. Severus, Galerius: *Paneg. Lat.* 12(9) 3,4; Lactantius, *De mort.* 26f. *Maiestas*: *Paneg. Lat.* 12(9) 3,7 and 15,1. For *comites*-horse guard detachments in Italy see Lactantius, *De mort.* 26,5f; Hoffmann, 1969/70, 244f; Speidel 1992, 377f. Galerius later used as horse guard units the *vexillationes palatinae* of the *promoti*, *clibanarii* and *sagittarii* (Lactantius, *De mort.* 40; *Not. Dig. Or.* 5,13).

202. Praetorians: *Paneg. Lat.* 12(9), 17; Durry 1938, 393. Praetorian horsemen (*equites promoti domnici*; Speidel 1992, 388f) no doubt also fought. Tactics: *Paneg. Lat.* 12(9)4: *necessitatem resistendi . . . propinquitatem refugiendi.* Onasander 33; see Herodotus 5,118,2. Mauri: Lactantius, *De mort.* 44; Arrian, *Parth.*, frag. 52 Roos; Herodian 6,7,5; Dio 78,27. Hinged on the horse: Zosimus 2,16. Maxentius' Italians: Zosimus 2,16,3. Treason in Rome: Lactantius, *De mort.* 44,7.

203. Constantine's Germans: Zosimus 2,15; huge-limbed etc: *Paneg. Lat.* 4(10),14. For the battle, see Speidel 1992, 279–85 (though the ripa ulterior surely was the south bank). Whirlpool: *Paneg. Lat.* 12(9),17; later accounts say that Maxentius fell from the bridge, but if so, he would not have had his horse to swim with as the Panegyrist says. Plate 20: photograph Deutsches Archologisches Institut, Rome, Inst. Neg. 32–74. *Clibanarii* with armored horses: *Paneg. Lat.* 4(10), 22; Speidel 1992, 406–13; compare Bishop-Coulston 1983, 182. Scale: *RIB* 201; CIL XIII 8095, etc; see above, p. 106; also the Trajanic battle relief on Constantine's arch. Galerius' foot and horse guard on his arch in Thessalonica (Laubscher 1975, 45–8) likewise wear scale.

204. Italians: Tacitus, *Hist.* 2,17; Herodian 2,11,3–5; see Vegetius 1,28. Fierce fighters beyond: Tacitus, *Ger.* 32 with Much's comment; Seneca, *Ep.* 36,7. When in 238 the Senators rebelled against Maximinus, they, too, had to bring German troops into the city: Herodian 8,6 and 8,8. Foreigners: Speidel 1992, 71–81. For the *iuventus* of the cities still being warlike see Forni 1992, 69–71; see also E. Birley, 1988, 394. Reformers: Vegetius, e.g. 1,5. For Dio on the upper class, see above, p.101. For the Roman reaction under Constantine against the barbarization of the army, see Lippold 1991, 572.

205. No hope for mercy: *Paneg. Lat.* 12(9),17. Killing ended: ibid. 20 (with Paschoud 1971, 89). Winners' ranks: *Paneg. Lat.* XII, 21,2,f (Constantine; see also Zosimus 2,15,1); Hoffmann 1969, vol. 2, p. 11, n 41. Transfer (*militiae mutatio*): Digest 49,16,3. Flatterer: *Paneg. Lat.* 12(9),21. *Catafractarii:* Hoffman 1969, 69ff; 196f; 296; 483f; on the Rhine: CIL XIII, 6238 = Schleiemacher 1984, 49.

206. *Pro re publica:* Speidel 1993, 756; Vegetius 2,5 and 2,14. Avenger: CIL VI, 1139 = Dessau 694. Praetorianism: Frank 1969, 232. Forts: Zosimus: 2,17,2. Archaeology: Colini 1944, 313ff and 343ff. Another flatterer: *Paneg. Lat.* 4(10),6. Graveyard: Deichmann-Tschira 1957, 69f; Guyon 1987, 30–3; 49; 237–8. Terrible punishment: Horace, *Ep.* 16,11ff. *Paganissimo:* Panciera 1974, 221; divine providence: Ferrua 1949, 661. Constantine's view: Eusebius, *HE* 9,9. The author of *Paneg. Lat.* 12, our best source, being a pagan (Barnes 1981, 46f), keeps to politics; for Constantine's official war propaganda see Creed 1984, 117. Christian horsemen: above, p. 143. Sign of Christ: Lactantius, *De mort.* 44, see Creed 1984, 119.

207. For Constantine's buildings over the graveyard, see Guyon 1987. Saints buried there: Deichmann-Tschira 1957, 73 and 76; Pietri 1976, 30; Guyon 1987; Seeliger 1987. Churches endowed: Seeliger 1987, 73. Sarcophagus: Pietri 1976, 32; Guyon 1987, 257. 'Queen of the battlefield': Durry 1938, 395; *scholae:* above, p. 75. Gravestones: Speidel 1993; (255 hitherto unpublished pieces); for a stretch of the wall with gravestones still embedded, see ibid. p.2.

Conclusion

208. *Scholae:* above, p. 75. Franks: Frank 1969, 59ff. Germans: Hoffmann 1969, 299f; Medieval aristocracy: Demant 1980.

BIBLIOGRAPHY

G. Alföldy	*Die Hilfstruppen der römischen Provinz Germania inferior* (Epigraphische Studien 6), Düsseldorf 1968.
G. Alföldy	*Die römischen Inschriften aus Tarraco*, Berlin 1975.
G. Alföldy	*Römische Heeresgeschichte, Beitrge 1962–1985* (*Mavors* 3), Amsterdam 1987.
G. Alföldy	*Die Krise des römischen Reiches* (*Habes* 5), Stuttgart 1989.
R. Ames	*Sun Tzu, The Art of Warfare*, New York 1993.
H. Ankersdorfer	*Studien zur Religion des römischen Heeres von Augustus bis Diokletian*, Dissertation, Konstanz 1973.
E.M. Ausbüttel	'Zur rechtlichen Lage der römischen Militärvereine', *Hermes* 113, 1985, 500–5.
D. Ayalon	'Notes on the Furūsiyya Exercises and Games in the Mamluk Sultanate'. Uriel Heyd (ed.), *Studies in Islamic History and Civilization*, Jerusalem 1961, 31–62.
D. Baatz	'Zur Geschützbewaffnung römischer Auxiliartruppen in der frühen und mittleren Kaiserzeit', *Bonner Jahrbücher* 166, 1986, 194–207.
D. Baatz	'Ein Katapult der Legio IV Macedonica aus Cremona', *Mitteilungen des Deutschen Archäologischen Instituts, Römische Abteilung*, 87, 1980, 283–99.
P.K. Baillie Reynolds	'The Troops Quartered in the Castra Peregrinorum', *Journal of Roman Studies* 13, 1923, 168–89.
P.K. Baillie Reynolds	*The Vigiles of Imperial Rome*, Oxford 1926.
J.P.V.D. Balsdon	*The Emperor Gaius (Caligula)*, Oxford 1934.
J.-Ch. Balty	'Apamée (1986): Nouvelles données sur l'armée romaine d'orient et les raids sassanides du milieu du IIIe siècle', *Comptes Rendus, Académie des Inscriptions & Belles-Lettres*, 1987, 213–42.
M. Bang	*Die Germanen im römischen Dienst* Berlin, 1906.
L. Barkóczi etc.	*Intercisa*, I, Budapest 1954.
T.D. Barnes	*Constantine and Eusebius*, Cambridge/Massachusetts 1981.
T.D. Barnes	*The New Empire of Diocletian and Constantine*, Cambridge/Massachusetts 1982.

A.A. Barrett	*Caligula, the Corruption of Power*, London 1989.
M.C. Bartusis	*The Late Byzantine Army*, Philadelphia 1992.
G. Bauchhenss	'Der Zwischensockel einer Jupitergigantensäule aus Weisenheim a. Sand', *Mitteilungen des Historischen Vereins der Pfalz*, 73, 1976, 167–74.
G. Bauchhenss & G. Neumann (eds)	*Matronen und verwandte Gottheiten (Beihefte der Bonner Jahrbücher* 44), Köln 1987.
J. Beaujeu	*La religion romaine à l'apogée de l'empire. I, La politique religieuse des Antonins (96–192)*, Paris 1955.
O. Behrends	*Die Rechtsregelungen der Militärdiplome und das die Soldaten des Prinzipats treffende Eheverbot.* Eck-Wolf 1986, 116–66.
H. Bellen	*Die germanische Leibwache der römischen Kaiser des iulisch-claudischen Hauses* (Akad. Wiss.u.Lit. Mainz, Abhandl. d. geistes- und sozialwiss. Kl. 1), Mainz, 1981.
N. Benseddik	*Les troupes auxiliaires de l'armée romaine en Maurétanie Césarienne sous le Haut-Empire*, Algiers 1982.
M.G. Bertinelli Angeli	'Gli effetivi della legione e della coorte pretoria e i latercoli dei soldati missi honesta missione', *Rendiconti, Istituto Lombardo, Lettere*, 108, 1974, 3–12.
A. Birley	*Marcus Aurelius: A Biography*, 2nd. ed., London 1987.
A. Birley	*The African Emperor, Septimius Severus*, London 1988.
A. Birley	*Officers of the Second Augustan Legion in Britain*, Caerleon 1990.
A. Birley	'A Review of the Tablets by Period' in: E.R. and A. Birley, *Vindolanda Research Reports* NS II, Hexham 1993, 18–72.
E. Birley	*Roman Britain and the Roman Army*, Kendal 1961.
E. Birley	*The Roman Army. Papers 1929–1986 (Mavors* 4), Amsterdam 1988.
R. Birley	*The Roman Documents from Vindolanda*, Carvoran 1990.
M.C. Bishop	'Cavalry Equipment of the Roman Army in the First Century AD' in: Coulston 1988, 67–195.
M.C. Bishop & J.C.N. Coulston	*Roman Military Equipment from the Punic Wars to the Fall of Rome*, London 1993.
J. Bogaers	'Civitas en stad van de Bataven en Canninefaten', *Berichten van de Rijksdienst voor het Oudheidkundig Bodemonderzoek* 10–11, 1960–61, 263–317.
J. Bogaers	'Civitates und Civitas-Hauptorte in der nördlichen Germania Inferior', *Bonner Jahrbücher* 172, 1972, 310–33.
W. Boppert	*Militärische Grabdenkmäler aus Mainz und Umgebung*

	(*Corpus Signorum Imperii Romani, Deutschland* II,3), Mainz 1992.
A.K. Bowman & J.D. Thomas	*Vindolanda, The Latin Writing Tablets*, London 1983.
D.J. Breeze	'The Organization of the Career Structure of the *Immunes* and *Principales* of the Roman Army', *Bonner Jahrbücher* 174, 1974, 245–92.
R. Brilliant	*The Arch of Septimius Severus in the Roman Forum*, Rome 1967.
G.M. Browne	*Documentary Papyri from the Michigan Collection*, Toronto 1970.
P.A. Brunt	'Conscription and Volunteering in the Roman Imperial Army', *Scripta Classica Israelica* 1, 1974, 90–115.
P.A. Brunt	'Did Imperial Rome Disarm her Subjects?', *Phoenix* 29, 1975, 260–70.
F. Bücheler	*Carmina latina epigraphica*, 2 vols. Leipzig 1895–1897.
R. Cagnat	*L'armée romaine d'Afrique*, Paris 1913.
D.B. Campbell	'Auxiliary Artillery Revisited', *Bonner Jahrbücher* 186, 1986, 117–32.
J.B. Campbell	*The Emperor and the Roman Army 31 BC–AD 235*, Oxford 1984.
D.G. Chandler	*The Campaigns of Napoleon*, New York 1966.
G.L. Cheesman	*The Auxilia of the Roman Imperial Army*, Oxford 1914.
M. Clauss	*Untersuchungen zu den Principales des römischen Heeres von Augustus bis Diocletian*, Dissertation Bochum, 1973(a).
M. Clauss	'Probleme der Lebensalterstatistiken aufgrund römischer Grabinschriften', *Chiron* 3, 1973(b), 395–417.
M. Clauss	'Frumentarius Augusti', *Epigraphica* 42, 1980, 131–4.
M. Clauss	*Cultores Mithrae. Die Anhängerschaft des Mithras-Kultes.* Stuttgart 1992.
A.M. Colini	*Storia e topografia del Celio nell'antichità*, Rome 1944.
H.M. Colton & J. Geiger	*Masada II, Final Reports. The Latin and Greek Documents*, Jerusalem 1989.
P. Connolly	'The Roman Saddle', in: Dawson 1987, 7–27.
P. Connolly	*Tiberius Claudius Maximus, the Cavalryman*, Oxford 1988.
P. Couissin	*Les armes romaines*, Paris 1926.
J.C. Coulston	'Roman Archery Equipment' in: M.C. Bishop, *The Production and Distribution of Roman Military Equipment*, Oxford 1985, 220–366.

J.C. Coulston	'Roman Military Equipment on Third Century Tombstones', in: Dawson 1987, 141–56.
J.C. Coulston (ed.)	*Military Equipment and the Identity of Roman Soldiers*, Oxford 1988 (=BAR Intl. 394).
J.L. Creed	*Lactantius, De Mortibus Persecutorum*, Oxford 1984.
F. Cumont	*Récherches sur le symbolisme funéraire des romains*, Paris 1942.
Curiosum	*Curiosum urbis Romae regionum XIV cum breviariis suis*, herausgegeben von O. Richter, *Topographie von Rom*, Leipzig 1901.
M. Dawson (ed.)	*Roman Military Equipment, The Accoutrements of War. Proceedings of the Third Roman Military Equipment Research Seminar*, Oxford 1987.
R.W. Davies	*Service in the Roman Army*, New York 1989.
J.G. Deckers, H.R. Seeliger & G. Mietke	*Die Katakombe 'Santi Marcellino e Pietro', Repertorium der Malereien*, Città del Vaticano and Münster 1987.
F.W. Deichmann & A. Tschira	'Das Mausoleum der Kaiserin Helena und die Basilika der Heiligen Marcellinus und Petrus an der Via Labicana vor Rom', *Archäologischer Anzeiger* 72, 1957, 44–110.
A. Demant	'Der spätrömische Militäradel', *Chiron* 10, 1980, 609–36.
G.T. Dennis	*Three Byzantine Military Treatises*, Washington DC 1985.
H. Dessau	*Inscriptiones Latinae Selectae.* Berlin 1892 ff.
D. Detschew	*Die thrakischen Sprachreste*, Vienna 1957.
H. Devijver	*The Equestrian Officers of the Roman Imperial Army I* (*Mavors* 6), Amsterdam 1989; II (*Mavors* 9), Stuttgart 1992.
H. Devijver	*Prosopographia militiarum equestrium*, 4 vols. Leuven 1976–1987.
J. de Vries	*Keltische Religion*, Stuttgart 1961.
B. Dobson	(see Domaszewski 1967)
B. Dobson	*Die Primipilares. Entwicklung und Bedeutung, Laufbahnen und Persönlichkeiten eines römischen Offiziersranges*, Bonn 1978.
B. Dobson & D.J. Breeze	'The Rome Cohorts and the Legionary Centurionates', *Epigraphische Studien* 8, 1969, 100–324.
A.v. Domaszewski	'Die Fahnen im römischen Heere', *Abhandlungen des Archäologisch-Epigraphischen Seminars der Universität Wien*, Heft V, 1885, 1–80. Reprint Darmstadt 1972.
A.v. Domaszewski	'Die Religion des römischen Heeres', *Westdeutsche Zeitschrift für Geschichte und Kunst* 14, 1895, 1–124. Reprint Darmstadt 1972.

206

A. v. Domaszewski	*Die Rangordnung des römischen Heeres*, Bonn 1908; 2nd ed., B. Dobson, Köln 1967.
C. van Driel-Murray (ed.)	*Roman Military Equipment: The Sources of Evidence*, Oxford 1989.
R. Duncan-Jones	*The Economy of the Roman Empire*, 2nd ed., Cambridge 1982.
M. Durry	'Juvénal et les Prétoriens', *Revue des Études Latines* 1935, 95ff.
M. Durry	*Les cohortes prétoriennes*, Paris 1938.
W. Eck & H. Wolff	*Heer und Integrationspolitik. Die römischen Militärdiplome als historische Quelle*, Köln-Vienna 1986.
B. Farwell	*Armies of the Raj. From the Great Indian Mutiny to Independence: 1858–1947*, New York 1989.
A. Ferrua	'Antichità Christiane. La guardia a cavallo', *Civiltà Cattolica* 1949, 523–31 und 654–61.
A. Ferrua	'Nuove iscrizioni degli equites singulares', *Epigraphica* 13, 1951, 96–141.
R. O. Fink	*Roman Military Records on Papyrus*, Cleveland/Ohio 1971.
K. Fittschen	'Ein Bildnis in Privatbesitz. Zum Realismus römischer Porträts der mittleren und späteren Prinzipatszeit'. *Eikones*, ed. P. A. Stucky, Bern 1980, 108–14.
K. Fittschen & P. Zanker	*Katalog der römischen Porträts in den Capitolinischen Museen und den anderen kommunalen Sammlungen der Stadt Rom*, I, Mainz 1985.
J. Fitz	*Honorific Titles of Roman Military Units in the 3rd Century*, Budapest-Bonn, 1983.
F. B. Florescu	*Die Trajanssäule, Grundfragen und Tafeln*, Bonn 1969.
G. Forni	*Il reclutamento delle legioni da Augusto a Diocleziano*, Milan 1953.
G. Forni	*Le tribù Romane, III,1, Le Pseudo-Tribù*, Rome 1985
G. Forni	*Esercito e Marina di Roma Antica* (*Mavors* 5), Stuttgart 1992.
R. I. Frank	*Scholae Palatinae*, Rome 1969.
C. Franzoni	*Habitus atque habitudo militis, Monumenti funerari di militari nella Cisalpina Romana*, Rome 1987.
F. Fröhlich	*Die Gardetruppen der römischen Republik*, Aarau 1882.
H. Gabelmann	'Ein Iuventusrelief in Fossombrone', *Festschrift für Nikolaus Himmelmann*, ed. H.-U. Cain, H. Gabelmann, D. Salzmann, Mainz 1989.
J. Garbsch	*Römische Paraderüstungen*, München 1979.
P. Garnsey & R. Saller	*The Roman Empire. Economy, Society and Culture*, Berkeley 1987.

W. Gauer	*Untersuchungen zur Trajanssäule*, Berlin 1977.
M. Gelzer	*Caesar der Politiker und Staatsmann*, Wiesbaden 1989.
G. Geraci	'La basilike ile macedone e l'esercito dei primi Tolemei', *Aegyptus* 69, 1979, 8–24.
B. Gerov	*Inscriptiones Latinae in Bulgaria Repertae*, Sofia 1989.
J.F. Gilliam	'Some Latin Military Papyri from Dura', *Yale Classical Studies* 11, 1950, 171–252.
J.F. Gilliam	*Roman Army Papers* (*Mavors* 2), Amsterdam 1986
A. Giuliano	*Arco di Costantino*, Milano 1955.
A. Giuliano	*Museo Nazionale Romano, Le Sculture I,7* (2 vols.), Rome 1984.
F. Gnecchi	*I Medaglioni Romani*, 3 vols. Milan 1912.
R. Göbl	*Der Triumph des Sasaniden Shapur über die Kaiser Gordianus, Philippus und Valerianus*, Wien 1974.
A.E. Gordon	*Illustrated Introduction to Latin Epigraphy*, Berkeley 1983.
M. Grant	*The Army of the Caesars*, New York 1974.
W.B. Griffiths	'The Sling and its Place in the Roman Imperial Army', in van Driel-Murray 1989, 253–79.
W. Groenman-van Wateringe	'Römische Lederfunde aus Vindonissa und Valkenburg Z.H., ein Vergleich', *Jahresbericht der Gesellschaft Pro Vindonissa* 1974, 62–84.
R. Grosse	*Römische Militärgeschichte von Gallienus bis zum Beginn der byzantinischen Themenverfassung*, Berlin 1920.
F. Grosso	*La lotta politica al tempo di Commodo*, Turin 1964.
F. Grosso	'Il diritto latino ai militari in età Flavia', *Rivista di cultura classica e medievale* 7, 1965(a), 541–60.
F. Grosso	'L'importanza dei corporis custodes nella successsione all'impero romano', *Clio* (Elsinore, Roma) 3, 1965(b) 389–406.
F. Grosso	'Equites singulares Augusti', *Latomus* 25, 1966, 900–9.
F. Grosso	'Tertulliano e l'uccisione di Pertinace', *Rendiconti morali della Acadenia dei Lincei*, Ser.8, vol. 21 (1966) 140–50.
F. Grosso	'Richerche su Plauziano e gli avvenimenti del suo tempo', *Accademia nazionale dei Lincei, Rendiconti* 23, 1968, 7–58.
J. Guyon	'Portica in Circuitu. Les annexes de la basilique constantinienne des saints Marcellin-et-Pierre sur la via Labicana à Rome,' in O. Feld and U. Peschlow, *Studien zur spätantiken und byzantinischen Kunst, I*, Mainz 1986, 235–48.
J. Guyon	*Le cimetière aux Deux Lauriers*, Rome 1987.

D. Hagedorn	'Marci Aurelii in Aegypten', *Bulletin of the American Society of Papyrologists* 16, 1979, 47–59.
H. Halfmann	*Itinera Principum, Geschichte und Typologie der Kaiserreisen im römischen Reich (Habes 2)*, Stuttgart 1986.
J.R. Hale	Renaissance War Studies, London 1983.
A. Hanson	'Private Letter', *Bulletin of the American Society of Papyrologists* 22, 1985, 87–96.
J. Harmand	*L'armée et le soldat à Rome de 107 à 50 avant notre ère*, Paris 1967.
W. Helbig	*Zur Geschichte der hasta donatica (=Abh. Wiss. Göttingen, Phil.-Hist. Kl. NF 10,3)*, Berlin 1908.
W. Helbig	*Führer durch die öffentlichen Sammlungen klassischer Altertümer in Rom*, I, Tübingen 1963; III, 1969.
P. Herz	'Die Ala Parthorum et Araborum', *Germania* 60, 1982, 173–182.
R.H. Hesselink	'The Introduction of the Art of Mounted Archery into Japan', *Transactions of the Asiatic Society of Japan* 6, 1991, 27–47.
O. Hirschfeld	*Kleine Schriften*, Berlin 1913.
D. Hoffmann	*Das spätrömische Bewegungsheer*, 2 vols., Düsseldorf 1969/70.
A. Holder	*Alt-Celtischer Sprachschatz*, 3 vols, Leipzig 1904.
P.A. Holder	*The Auxilia from Augustus to Trajan*, Oxford 1980.
L. Homo	*Essai sur le règne de l'empereur Aurélien (270–275)*, Paris 1904.
G. Horsmann	*Untersuchungen zur militärischen Ausbildung im republikanischen und kaiserzeitlichen Rom*, Boppard 1991.
Hyginus	See Lenoir 1979.
A. Hyland	*Equus: The Horse in the Roman World*, London 1990.
L. Jalabert, R. Mouterde & C. Mondésert	*Inscriptions grecques et latines de la Syrie*, Vol. 4, *Laodicène et Apamène*, Paris 1955.
S. James	'Dura-Europos and the Introduction of the "Mongolian Release"', in Dawson 1987, 77–83.
S. James	'*Fabricae*: State Arms Factories of the Later Roman Empire', in Coulston 1988, 257–331.
A. Johnson	*Römische Kastelle des 1. und 2. Jahrhunderts n. Chr. in Britannien und in den germanischen Provinzen des Römerreiches, bearbeitet von D. Baatz*, Mainz 1987.
E. Josi	'Scoperte nella basilica constantiniana al Laterano', *Rivista di Archeologia Cristiana* 1934, 335–358.
E. Josi	'Scoperte nella navata centrale della Basilica Lateranense', *Bolletino dell' istituto nazionale di archeologia e storia dell'arte* 6, 1933–36, 196–7.

M. Junkelmann	*Römische Kavallerie – Equites Alae*, Aalen 1989.
M. Junkelmann	*Die Reiter Roms, II Der militärische Einsatz (Kulturgeschichte der Alten Welt 49)*, Mainz 1991.
I. Kajanto	*The Latin Cognomina*, Helsinki 1965.
John Keegan	*The Face of Battle. A Study of Agincourt, Waterloo and the Somme*, London 1976.
D.L. Kennedy	'Some Observations on the Praetorian Guard', *Ancient Society* 9, 1978, 275–301.
D.L. Kennedy	'The Military Contribution of Syria to the Roman Imperial Army', French – Lightfoot 1989, 235–46.
D.L. Kennedy & D. Riley	*Rome's Desert Frontier from the Air*, London 1990.
E. Kettenhofen	*Die römisch-persischen Kriege des 3. Jahrhunderts n. Chr.*, Wiesbaden 1982.
F. Kiechle	'Die 'Taktik' des Flavius Arrianus', *Bericht der römisch-germanischen Kommission* 45, 1964, 87–129.
D. Kienast	*Untersuchungen zu den Kriegsflotten der römischen Kaiserzeit* (=*Antiquitas* 1/13), Bonn 1966.
G. Koch & H. Sichtermann	*Römische Sarkophage*, München 1982.
F. Kolb	*Literarische Beziehungen zwischen Cassius Dio, Herodian und der Historia Augusta*, Bonn 1972.
K. Kraft	*Zur Rekrutierung der Alen und Kohorten an Rhein und Donau*, Bern 1951.
K. Kraft	*Gesammelte Aufsätze zur antiken Geschichte und Militärgeschichte*, Darmstadt 1973.
R. Krautheimer, S. Corbett & A. Fraser	*Corpus Basilicarum* V, Rome 1977.
J. Krohmayer & G. Veith	*Heerwesen und Kriegführung der Griechen und Römer*, München 1928.
Th. Kraus	*Das römische Weltreich*, Berlin 1967.
W. Kubitschek	'Signifer', *RE* II,2 (1923) 2348–2358.
R. Lanciani	'Gli allogiamenti degli equites singulares', *Bulletino di archeologia communale* 1885, 137–56.
R. Lanciani	*New Tales of Old Rome*, 1901; Reprint New York 1967.
J.-M. Lassère	*Ubique Populus, Peuplement et mouvements de population dans l'Afrique romaine*, Paris 1977.
J.-M. Lassère	'Les Afri et l'armée romaine', *L'Africa Romana* 5, 1987, 177–188.
K. Latte	*Römische Religionsgeschichte*, München 1960.
H.P. Laubscher	*Der Reliefschmuck des Galeriusbogens in Thessaloniki*, Berlin 1975.
Y. Le Bohec	La troisième légion Auguste, Paris 1989a.
Y. Le Bohec	*L'armée romaine sous le haut-empire*, Paris 1989b.
Y. Le Bohec	*Les unités auxiliaires de l'armée romaine en Afrique Proconsulaire et Numidie sous le Haut Empire*, Paris 1989c.

M. Leglay	'Hadrien et Viator sur les champs de manoeuvre de Numidie', *Mélanges d'histoire ancienne offerts à William Seston*, Paris 1974, 277–83.
M. Lenoir	*Pseudo-Hygin, Des fortifications du camp*, Paris 1979.
F. Lepper & S. Frere	*Trajan's Column*, Glouchester 1988.
P. LeRoux	'L'amphithéâtre et le soldat sous l'empire romain', *Spectacula* I, *Gladiateurs et amphithéatres*, Laltes 1990, 203–15.
J. Lesquier	*L'armée romaine d'Égypte d'Auguste à Dioclétien* (=MIFAO 41), Cairo 1918.
H. Lieb	'Die Constitutiones für die stadtrömischen Truppen', in Eck und Wolff (1986) 322–46.
W. Liebenam	'Equites singulares', *RE* VI, 1907, 312–21.
A. Lippold	*Kommentar zur Vita Maximini Duo der Historia Augusta*, Bonn 1991.X.
Loriot	'Les premières années de la grande crise du IIIe siècle: De l'avènement de Maximin le Thrace (235) à la mort de Gordien III (244),' *ANRW* II, 2, 1975, 657–787.
D.D. Luckenbill	*Ancient Records of Assyria and Babylonia*, II, Chicago 1927.
A.A. Lund	'Gesamtinterpretation der Germania des Tacitus', *ANRW* II, 33/3, 1858–1988.
A.A. Lund	'Kritischer Forschungsbericht zur *Germania* des Tacitus', *ANRW* II, 33/3, 1989–2222.
R. MacMullen	'The Epigraphic Habit in the Roman Empire', *American Journal of Philology* 103, 1982, 233–46.
R. MacMullen	'Personal Power in the Roman Empire', *American Journal of Philology* 107, 1986, 512–24.
R. MacMullen	*Corruption and the Decline of Rome*, New Haven 1988.
R. MacMullen	*Changes in the Roman Empire, Essays in the Ordinary*, Princeton 1990.
A. Magnen	*Epona*, Paris 1953.
J.C. Mann	*Legionary Recruitment and Veteran Settlement during the Principate*, London 1983.
J.C. Mann	'The Castra Peregrina and the "Peregrini"', *Zeitschrift für Papyrologie und Epigraphik* 74, 1988, 148.
Ph. Mansel	*Pillars of Monarchy. An Outline of the Political and Social History of Royal Guards 1400–1984*, London 1984.
Chr. Marek	'Stadt, Ära und Territorium in Pontus-Bithynia und Nord-Galatia', *Istanbuler Forschungen* 39, Tübingen 1993.
R. Marichal	see *ChLA*
R. Marichal	*Les ostraca de Bu Njem*, Tripoli 1992.

V.A. Maxfield	*The Military Decorations of the Roman Army*, Berkeley 1981.
Maurice	*Strategikon*, ed. G.T. Dennis, Vienna 1980.
C. Mercurelli	'Il sarcofago di un centurione pretoriano cristiano e la diffusione del cristianesimo nelle coorti pretorie', *Rivista di Archeologia Cristiana* 16, 1939, 73–99.
F. Millar	*A Study of Cassius Dio*, Oxford 1964.
F. Millar	*The Emperor in the Roman World*, Ithaca NY, 1977.
M. Mirković	'Die römische Soldatenehe und der 'Soldatenstand'', *Zeitschrift für Papyrologie und Epigraphik 40*, 1980, 259–72.
A. Mócsy	*Pannonia and Upper Moesia*, London 1974.
A. Mócsy	Beiträge zur Namensstatistik, Budapest 1985.
A. Mócsy	*Pannonien und das römische Heer. Ausgewählte Aufsätze.* (= *Mavors* 7), Stuttgart 1992.
Th. Mommsen	*Römisches Staatsrecht*, 3rd ed., 3 vols., Leipzig 1885–1887.
Th. Mommsen	*Gesammelte Schriften*, 5, 6, 8, Berlin 1910 and 1913.
E.S. Morse	'Ancient and Modern Methods of Arrow-Release', *Bulletin of the Essex Institute* 17, 1885, 3–56.
R. Much	*Die Germania des Tacitus*, 3rd. ed. Heidelberg 1967.
E. Nash	*Pictorial Dictionary of Ancient Rome*, 2nd ed., New York 1968.
H. Nesselhauf & Hans Lieb	'Dritter Nachtrag zu *CIL* XIII', *Berichte der RGK* 40, 1959, 120–229.
M.P. Nilsson	*Geschichte der griechischen Religion* II, München 1974.
P. Noelke	'Ein neuer Soldatengrabstein aus Köln', *Studien zu den Militärgrenzen Roms* III, Stuttgart 1986, 213–26.
J. Nollé	'Side. Zur Geschichte einer kleinasiatischen Stadt in der römischen Kaiserzeit im Spiegel ihrer Münzen', *Antike Welt* 21, 1990, 244–65.
J. Oldenstein	'Zur Ausrüstung römischer Auxiliareinheiten', *Berichte RGK* 57, 1976, 44–284.
M. Pallottino	*Il grande fregio di Traiano*, Rome 1938.
S. Panciera	'Equites Singulares, Nuove testimonianze epigraphiche,' *Rivista di Archeologia Cristiana* 50, 1974, 221–47.
S. Panciera	'Silvano a Roma', Festschrift G. Mihailov, Sofia 1989 (in press).
S. Panciera	'Genio castrorum peregrinorum', *A. Arch. Hung.* 41, 1989, 365–83
F. Paschoud	*Zosime, Histoire nouvelle* I, Paris 1971.
A. Passerini	*Le coorti pretorie*, Rome 1939.
A. Passerini	'Legio' in: E. de Ruggiero, *Dizionario Epigraphico*, vol. 4, Rome 1950, 549–624.

E. Petersen, A.v. Domaszewski & G. Calderini	*Die Marcus-Säule auf der Piazza Colonna in Rom*, München 1896.
H.v. Petrikovits	*Die Innenbauten römischer Legionslager*, Opladen 1975.
H.v. Petrikovits	'Matronen und verwandte Gottheiten' in Bauchhenss-Neumann 1987, 241–54.
H.G. Pflaum	*Les carrières procuratoriennes équestres sous le haut-empire romain*, 3 vols, Paris 1960–61.
H.G. Pflaum	'Comes divini lateris', *BCTH*, NS 10–11, 1974–5, 196.
H.G. Pflaum	*Les carrières procuratoriennes équestres sous le haut-empire romain*, supplément, Paris 1982.
E. Pfuhl & H. Möbius	*Die ostgriechischen Grabreliefs* I, Mainz 1977.
Ch. Picard	'Sabazios, dieu thraco-phrygien: expansion et aspects nouveaux de son culte', *Revue Archéologique* 1961, 129–76.
J. Piekalkiewicz	*The Cavalry of World War II*, New York 1980.
Ch. Pietri	*Roma Christiana*, Rome 1976
L.F. Pitts & J.K. St. Joseph	*Inchtuthil, The Roman Legionary Fortress*, London 1985.
J. Pokorny	*Indogermanisches etymologisches Wörterbuch*, Bern 1959
K. Raaflaub	'Die Militärreformen des Augustus und die politische Problematik des frühen Prinzipats', in G. Binder (ed.) *Saeculum Augustum* I, Darmstadt 1987, 246–307.
N.B. Rankov	'M. Oclatinius Adventus in Britain', *Britannia* 18, 1987, 243–9.
G. Ravegnani	*Soldati di Bisanzio in età Giustiniana*, Rome 1988.
J.R. Rea	'Troops for Mauretania', *Zeitschrift für Papyrologie und Epigraphik* 26, 1977, 223–7.
J.R. Rea	'A Cavalryman's Career, AD 384(?)–401', *Zeitschrift für Papyrologie und Epigraphik* 56, 1984, 79–88.
W. van Rengen & J.Ch. Balty	*Apamea in Syria, Winterquarters of legio II Parthica*, Brussels 1992.
M. Reddé	*Mare Nostrum*, Rome 1986.
H. Reichert	*Lexikon der altgermanischen Namen*, Vienna 1987.
S. Reinach	'Epona' in *Revue Archéologique* 1895, 163–95 and 310–35.
J.-C. Richard	'Les aspects militaires des funérailles impériales', *École Française de Rome, Mélanges d'Archéologie* 78, 1966, 313–25.
E. Ritterling	'Ein Amtsabzeichen der beneficiarii consularis im Museum zu Wiesbaden', *Bonner Jahrbücher* 25, 1919, 9–34.

213

E. Ritterling	'Legio', *RE* XII, 1924–25, 1211–1829.
L. Robert	*Noms indigènes dans l'Asie Mineure gréco-romaine*, Paris 1963.
L. Robert	*A travers l'Asie mineure*, Paris 1980.
H.R. Robinson	*The Armour of Imperial Rome*, London 1975.
L. Rocchetti	'Su una stele del periodo tetrarchico', *Annuario della Scuola Archeologica Italiana di Atene*, 45–46, 1967–68, 487–498.
M.I. Rostovtzeff	*The Excavations at Dura-Europos, Preliminary Report of the Sixth Season*, New Haven 1933.
M.I. Rostovtzeff	*The Social and Economic History of the Roman Empire*, second edition, Oxford 1957.
M.M. Roxan	*Roman Military Diplomas 1954–1977*, London 1978.
M.M. Roxan	'The Distribution of Roman Military Diplomas', *Epigraphische Studien* 12, 1981, 265–86.
M.M. Roxan	*Roman Military Diplomas 1978–1984*, London 1985.
M. Roxan	'Findspots of Military Diplomas of the Roman Auxiliary Army', London Institute for Archaeology, *Bulletin* 26, 1990, 127–130.
M.M. Roxan	'Women on the Frontiers', *Roman Frontier Studies 1989*, ed. V.A. Maxfield and M.J. Dobson, Exeter 1991, 662–7.
C.B. Rüger	*Germania Inferior. Untersuchungen zur Territorial- und Verwaltungsgeschichte Niedergermaniens in der Prinzipatszeit*, Köln 1968.
I.S. Ryberg	*Panel Reliefs of Marcus Aurelius*, New York 1967.
D.B. Saddington	*The Development of the Roman Auxiliary Forces from Caesar to Vespasian (49 BC–AD 79)*, Harare 1982.
D.B. Saddington	'St. Ignatius, Leopards, and the Roman Army', *Journal of Theological Studies* 1987, 411f.
R.P. Saller & B.D. Shaw	'Tombstones and Roman Family Relations in the Principate: Civilians, Soldiers and Slaves', *Journal of Roman Studies* 74, 1984, 124–56.
J. Sasel	'Bellum Serdicense' *Situla* 4, 1961, 3–30.
M. Sasel	'A Latin Epitaph of a Roman Legionary from Corinth' *Journal of Roman Studies* 68, 1978, 22–5.
R. Saxer	*Untersuchungen zu den Vexillationen des römischen Kaiserheeres von Augustus bis Diokletian*, Köln 1967.
M. Sayar	'Equites Singulares Augusti in neuen Inschriften aus Anazarbos', *Epigraphica Anatolica* 17, 1991, 19–40.
Th. Schäfer	'Zum Schlachtsarkophag Borghese', *Mélanges de l'École Française de Rome*, 91, 1979, 355–70.
W. Scheidel	'Inschriftenstatistik und die Frage des Rekrutierungsalters römischer Soldaten', *Chiron* 22, 1992, 281–97.
U. Schillinger-Häfele	'Vierter Nachtrag zu *CIL* XIII', *Berichte der römisch-germanischen Kommission* 58, 1977, 448–603.

M. Schleiermacher	*Römische Reitergrabsteine. Die kaiserzeitlichen Reliefs des triumphierenden Reiters*, Bonn 1984.
A. Schober	*Die römischen Grabsteine von Noricum und Pannonien*, Vienna 1923.
H.R. Seeliger	'Die Geschichte der Katakombe 'Inter Duas Lauros' nach den schriftlichen Quellen', in Deckers usw. 1987, 59–90.
H.G. Simon	'Die Reform der Reiterei unter Kaiser Gallien', *Studien zur antiken Sozialgeschichte, Festschrift Vittinghoff*, ed. W. Eck *et al.*, Köln 1980, 435–52.
F. Sinn	*Stadtrömische Marmorurnen*, Mainz 1987.
J. Smeesters	'Les Tungri dans l'armée romaine', *Studien zu den Militärgrenzen Roms*, II, ed. D. Haupt, Köln 1977, 175–86.
C.S. Sommer	'Kastellvicus und Kastell', *Fundberichte aus Baden-Württemberg* 13, 1988, 457–707.
D. Sourdel	*Les cultes du Hauran à l'époque romaine*, Paris 1952.
P. Southern	'The Numeri of the Roman Imperial Army', *Britannia* 20, 1989, 81–140.
M.A. Speidel	'Entlassungsurkunden des römischen Heeres', *Gesellschaft Pro Vindonissa, Jahresbericht* 1990, 59–65.
M.A. Speidel	'Roman Army Pay Scales', *Journal of Roman Studies* 82, 1992, 87–106.
M.P. Speidel	*Die Equites Singulares Augusti. Begleittruppe der römischen Kaiser des zweiten und dritten Jahrhunderts*, Bonn 1965.
M.P. Speidel	*Guards of the Roman Armies. An Essay on the Singulares of the Provinces*, Bonn 1978a.
M.P. Speidel	*The Religion of Iuppiter Dolichenus in the Roman Army*, Leiden 1978b.
M.P. Speidel	*Mithras-Orion, Greek Hero and Roman Army God* (=*EPRO* 81), Leiden 1980.
M.P. Speidel	'Raetien als Herkunftsgebiet der kaiserlichen Gardereiter', *Bayerische Vorgeschichtsblätter* 46, 1981, 105–20 and 48, 1983, 187–8.
M.P. Speidel	'Noricum als Herkunftsgebiet der kaiserlichen Gardereiter', *Jahreshefte des österreichischen archäologischen Institutes* 53, 1982, 214–44.
M.P. Speidel	*Roman Army Studies* I (*Mavors* 1), Amsterdam 1984.
M.P. Speidel	'The Early Protectores and their Beneficiarius Lance', *Archäologisches Korrespondenzblatt* 16, 1986, 451–54.
M.P. Speidel	'Swimming the Danube under Hadrian's Eyes. A Feat of the Emperor's Batavi Horse Guard', *Ancient Society* 22, 1991, 277–82.
M.P. Speidel	*Roman Army Studies* II (*Mavors* 8), Stuttgart 1992.

M.P. Speidel	*The Framework of an Imperial Legion*, Caerleon 1992a.
M.P. Speidel	'Ala Celerum Philippiana', *Tyche* 1992b.
M.P. Speidel	*Die Denkmäler der Kaiserreiter (Equites singulares Augusti)* Bonn 1993.
M.P. Speidel	'Commodus the God-Emperor and the Army', *JRS* 1993a.
M.P. Speidel	'The *fustis* as a Soldier's Weapon', *Antiquités Africaines* 26, 1993b, 135–47.
M.P. Speidel	'Mauri Equites. The Tactics of Light Cavalry in Mauretania', *Antiquités Africaines* 26, 1993c, 119–24.
M.P. Speidel	'Legionary Horsemen as Guards' (forthcoming) 1994.
C.G. Starr	*The Roman Imperial Navy*, 2nd ed., New York 1960.
A. Stein	*Der römische Ritterstand*, München 1927.
E. Stein	*Die kaiserlichen Beamten und Truppenkörper im römischen Deutschland unter dem Prinzipat*, Vienna 1932.
E. Stein	*Histoire du Bas-Empire* I, Paris 1959.
B.H. Stolte	Die religiösen Verhältnisse in Niedergermanien, *ANRW* II, 18/1, 1986, 591–671.
K. Strobel	*Untersuchungen zu den Dakerkriegen Trajans*, Bonn 1984.
K. Strobel	'Anmerkungen zur Geschichte der Bataverkohorten in der hohen Kaiserzeit', *Zeitschrift für Papyrologie und Epigraphik* 70, 1987, 271–92.
K. Strobel	'Zu Fragen der frühen Geschichte der römischen Provinz Arabia, etc', *Zeitschrift für Papyrologie und Epigraphik* 71, 1988, 251–80.
K. Strobel	*Die Donaukriege Domitians*, Bonn 1989.
T. Sulimirski	*The Sarmatians*, New York.
R. Syme	*Tacitus*, Oxford 1958.
R. Syme	*Emperors and Biography*, Oxford 1971.
R. Syme	*Roman Papers* III and VI, Oxford 1984, 1991.
W.W. Tarn	*Alexander the Great*, Boston 1956.
B.E. Thomasson	*Laterculi Praesidum*. Vol. 1, Göteborg 1984.
D. Timpe	*Arminius-Studien*, Heidelberg 1970.
J.M.C. Toynbee	*Animals in Roman Life and Art*, Ithaca/NY 1973.
H. Ubl	*Waffen und Uniform des römischen Heeres der Prinzipatsepoche nach den Grabreliefs Noricums und Pannoniens*, Ungedruckte Dissertation, Vienna 1969.
Z. Visy	'Die Entlassung der Auxiliarsoldaten aufgrund der Militärdiplome', *Acta Archaeologica Hungarica* 36, 1984, 223–38.

Th. Völling	'Funditores im römischen Heer', *Saalburg Jahrbuch* 45, 1990, 24–58.
E. Vorbeck	*Militärinschriften aus Carnuntum*, 2nd ed. Vienna 1980.
W. Wagner	*Die Dislokation der römischen Auxiliarformationen in den Provinzen Norikum, Pannonien, Moesien und Dakien von Augustus bis Gallienus*, Berlin, 1938.
R.E. Walker	'Some Notes on Cavalry Horses in the Roman Army', in Toynbee 1973, 335–43.
G. Walser	*Rom, das Reich, und die fremden Völker in der Geschichtsschreibung der frühen Kaiserzeit*, Baden-Baden 1951.
G. Walser	'Kaiser Domitian in Mainz', *Chiron* 19, 1989, 449–56.
G. Webster	*The Roman Imperial Army*, 3rd ed., London 1985.
E.L. Wheeler	'The occasion of Arrian's *Tactica*', *Greek, Roman and Byzantine Studies*, 19, 1978, 351–65.
C.R. Whittaker	*Herodian* (Loeb), Cambridge/Mass. 1969/70.
L. Wierschowski	*Heer und Wirtschaft. Das römische Heer der Prinzipatszeit als Wirtschaftsfaktor*, Bonn 1984.
W. Will	'Römische 'Klientel-Randstaaten' am Rhein?', *Bonner Jahrbücher* 187, 1987, 1–61.
W.J.H. Willems	'Romans and Batavians. A Regional Study in the Dutch Eastern River Area, II', *Berichten van de Rijksdienst voor het Oudheidkundig Bodemonderzoek* 34, 1984, 39–207.
H. Wolff	'Die cohors II Tungrorum milliaria equitata c(oram?) l(audata?) und die Rechtsform des ius Latii', *Chiron* 6, 1976, 267–88.
H. Wolff	'Das Fragment eines Auxiliardiploms von 203 n.Chr. in der Münzlehrsammlung der Universität Passau', *Ostbairische Grenzmarken, Passauer Jahrbuch* 29, 1987, 33–41.
R. Wormser	*The Yellowlegs. The Story of the United States Cavalry*. Garden City, New York, 1966.
R. Ziegler	*Städtisches Prestige und kaiserliche Politik. Studien zum Festwesen in Ostkilikien im 2. und 3. Jahrhundert n. Chr.* Düsseldorf 1985.
W. Zwikker	'Bemerkungen zu den römischen Heeresfahnen in der älteren Kaiserzeit', Bericht *RGK* 27, 1937, 7–22.

INDEX

For main discussions see hyphenated page numbers